BOLD VENTURES

Volume 1

Patterns Among U.S. Innovations in Science and Mathematics Education

W0051422

Other Volumes in the Series: Bold Ventures

Volume 2: Case Studies of U.S. Innovations in Science Education (ISBN 0-7923-4232-1)

Building on Strength: Changing Science Teaching in California Public Schools
J. Myron Atkin, Jenifer V. Helms, Gerald L. Rosiek, Suzanne A. Siner

The Different Worlds of Project 2061
J. Myron Atkin, Julie A. Bianchini, Nicole I. Holthuis

The Challenges of Bringing the Kids Network to the Classroom
James W. Karlan, Michael Huberman, Sally H. Middlebrooks

Science, Technology, and Story: Implementing the Voyage of the Mimi
Sally H. Middlebrooks, Michael Huberman, James W. Karlan

ChemCom's Evolution: Development, Spread, and Adaptation
Mary Budd Rowe, Julie E. Montgomery, Michael J. Midling, Thomas M. Keating

Volume 3: Case Studies of U.S. Innovations in Mathematics Education (ISBN 0-7923-4233-X)

Setting the Standards: NCTM's Role in the Reform of Mathematics Education
Douglas B. McLeod, Robert E. Stake, Bonnie P. Schappelle,
Melissa Mellissinos, Mark J. Gierl

Teaching and Learning Cross-Country Mathematics: A Story of Innovation in Precalculus
Jeremy Kilpatrick, Lynn Hancock, Denise S. Mewborn, Lynn Stallings

The Urban Mathematics Collaborative Project: A Study of Teacher, Community, and Reform
Norman L. Webb, Daniel J. Heck, William F. Tate

BOLD VENTURES

Volume 1

Patterns Among U.S.
Innovations in Science and Mathematics
Education

edited by

Senta A. Raizen
Edward D. Britton

from

**The National Center for Improving
Science Education**

a division of
The NETWORK, Inc.

SPRINGER-SCIENCE+BUSINESS MEDIA, B.V.

A C.I.P. Catalogue record for this book is available from the Library of Congress.

ISBN 978-0-7923-4235-9   ISBN 978-94-011-5440-6 (eBook)
DOI 10.1007/978-94-011-5440-6

Printed on acid-free paper

The National Center for Improving Science Education

The National Center for Improving Science Education (NCISE) is a division of The NETWORK, Inc., a nonprofit organization dedicated to educational reform. The Center's mission is to promote change in state and local policies and practices in science curriculum, teaching, and assessment. To further this mission, we carry out research, evaluation, and technical assistance. Based on this work, we provide a range of products and services to educational policymakers and practitioners to help them strengthen science teaching and learning across the country.

We are dedicated to helping all stakeholders in science education reform, preschool to postsecondary, to promote better science education for all students.

Advisory Board, U.S. Case Studies

Contents

Contributors

The Case Study Teams

Studies in Science Education

Building on Strength: Changing Science Teaching in California Public Schools
J. Myron Atkin, Jenifer V. Helms, Gerald L. Rosiek, Suzanne A. Siner
School of Education, Stanford University

The Different Worlds of Project 2061
J. Myron Atkin, Julie A. Bianchini, Nicole I. Holthuis
School of Education, Stanford University

The Challenges of Bringing the Kids Network to the Classroom
James W. Karlan, Michael Huberman, Sally H. Middlebrooks
National Center for Improving Science Education, Harvard University

Science, Technology, and Story: Implementing the Voyage of the Mimi
Sally H. Middlebrooks, Michael Huberman, James W. Karlan
National Center for Improving Science Education, Harvard University

ChemCom's Evolution: Development, Spread, and Adaptation
Mary Budd Rowe, Julie E. Montgomery, Michael J. Midling, Thomas M. Keating
Stanford University

Studies in Mathematics Education

Setting the Standards: NCTM's Role in the Reform of Mathematics Education
Douglas B. McLeod, Bonnie P. Schappelle, Melissa Mellissinos
San Diego State University

Robert E. Stake, Mark J. Gierl
University of Illinois at Champaign-Urbana

Teaching and Learning Cross-Country Mathematics: A Story of Innovation in Precalculus
Jeremy Kilpatrick, Lynn Hancock, Denise S. Mewborn, Lynn Stallings
University of Georgia

The Urban Mathematics Collaborative Project: A Study of Teacher, Community, and Reform
Norman L. Webb, Daniel J. Heck, William F. Tate
University of Wisconsin-Madison

The Editors

Senta A. Raizen, Director of The National Center for Improving Science Education, is principal investigator and editor of the U.S. case studies discussed in this and two companion volumes. Raizen is the primary author of a number of books, reports, and articles on science education in elementary, middle, and high school; indicators in science education; preservice education of elementary school teachers; and technology education. Her work also includes educational assessment and program evaluation, education policy, reforming education for work, and linking education research and policy with practice. She is principal investigator for NCISE research for the Third International Mathematics and Science Study (TIMSS) and serves on the TIMSS International Steering Committee. Raizen directs NCISE evaluations of several federal programs that support science education. She serves in an advisory capacity to—among others—the National Assessment of Educational Progress, the National Goals Panel, the National Institute for Science Education, and the National Research Council.

Edward D. Britton, Associate Director of NCISE, serves as project director for several international studies, including the work presented in this volume. He was lead editor of *Examining the Examinations: An International Comparison of Science and Mathematics Examinations for College-Bound Students*. Britton also works on several aspects of TIMSS, including the U.S. and international curriculum analyses and the international teacher and student questionnaires. In addition, he has managed development of CD-ROM disks and videotapes designed to help elementary teachers enhance their science knowledge and pedagogy. Britton has written on indicators for science education, dissemination of innovations, and evaluation.

Senior Authors

J. Myron Atkin is a professor of education at Stanford University and served as dean of education from 1979 to 1986. He coedited *Changing the Subject*, the Organisation for Economic Co-operation and Development's 1996 report on member nations' case studies of innovations in science, mathematics, and technology education. His research interests and publications focus on identification of the science content to be taught in elementary and secondary schools; teacher-initiated inquiry, especially action research; practical reasoning by teachers and children; case methods in educational research; evaluation of educational programs; science education in museums; and development of policies that accord classroom teachers greater influence in determining the educational research agenda.

Michael Huberman, formerly a professor of education at the University of Geneva (1972-93) and now professor emeritus, is presently director of research at the Swiss Federal Institute of Professional Education and, since 1991, visiting professor of education at Harvard University. He also was a senior researcher at the National Center for Improving Science Education where he led two case studies in this project. His areas of interest are qualitative research, research use, longitudinal studies of teaching, and educational innovation. He is a coauthor, with Matthew Miles, of *Qualitative data analysis: A sourcebook of new methods*, and also recently published *The Lives of Teachers*.

Jeremy Kilpatrick is a Regents Professor of Mathematics Education at the University of Georgia. For the Precalculus case study, Kilpatrick was the primary researcher at the Eisenhower High School site and was one of the primary researchers at the North Carolina School of Science and Mathematics. Kilpatrick's recent professional activities include work with the International Commission on Mathematical Instruction, the Bacomet Group, the National Council of Teachers of Mathematics (NCTM), and the College Board. Kilpatrick also serves as a board member or reviewer for a number of national and international journals.

Douglas B. McLeod has been professor of mathematics at San Diego State University since 1972. He also served from 1979 to 1981 as a program manager at the National Science Foundation and as a professor of mathematics and education at Washington State University from 1986 to 1993. His Ph.D. is from the Department of Mathematics at the University of Wisconsin. Previous research projects have focused on mathematics teaching and learning, with a special emphasis on affective issues in mathematics education. Within the National Council of Teachers of Mathematics, he has served on several committees, including 10 years on the editorial board of the *Journal for Research in Mathematics Education*.

Mary Budd Rowe was professor of science education at Stanford University and the University of Florida. Her career-long interest in finding ways to have all students experience and understand science in part prompted her to lead the case study of ChemCom, a course seeking to bring chemistry to more students. Rowe's leading-edge work brought her many tributes, including the annual *Journal of Research in Science Teaching* award for her seminal paper that coined the term "wait time," and the National Science Teachers Association's (NSTA's) Roger Carleton Award for national leadership in science education. She later was elected president of NSTA. In her last decade, Rowe focused particularly on pressing new technologies into the service of professional development for science teachers, including videotapes, CD-ROM disks, and telecommunication networks.

Robert E. Stake is a professor of education and director of the Center for Instructional Research and Curriculum Evaluation at the University of Illinois. Among the evaluative studies he has directed are works in science and mathematics in elementary and secondary schools, model programs and conventional teaching of the arts in schools, development of teaching with sensitivity to gender equity, education of teachers for the deaf and for youth in transition from school to work settings, environmental education and special programs for gifted students, and the reform of urban education. Among his writings are *Quieting Reform*, a book on Charles Murray's evaluation of Cities-in-Schools; *Custom and Cherishing*, a book with Liora Bresler and Linda Mabry on teaching the arts in ordinary elementary school classrooms in America; and *The Art of Case Study Research*, a book on research methods.

Norman L. Webb is a professor of mathematics at the University of Wisconsin-Madison. He served on the Urban Mathematics Collaborative Documentation Project that recorded, analyzed, and described the development of the collaboratives from their initial funding by the Ford Foundation. Webb's research focuses on mathematics and mathematics education, with a major emphasis on assessment and evaluation. He was part of the writing team of the NCTM *Curriculum and Evaluation Standards for School Mathematics* (1989).

Preface

This book, based on detailed studies of eight innovations in mathematics and science education, has many insights to offer on current school reform. Since each innovation studied has taken its own unique approach, the set as a whole spans the spectrum from curriculum development to systemic reform, from concentrating on particular school populations to addressing all of K-12 education. Yet these reform projects share a common context, a world view on what matters in science and mathematics for students of the 1990s and beyond, convictions about what constitutes effective instruction, and some notions about how school change can be brought about. These commonalities are drawn out in the book and illustrated with examples from the individual case studies that are reported in full in *Bold Ventures*, Volumes 2 and 3.

The eight innovations—all of them projects that are well-known, at least by name, to U.S. audiences—are briefly described in chapter 1. Each was the subject of an in-depth, three-year case study. The research teams analyzed many documents, attended numerous project meetings, visited multiple sites, conducted dozens of individual interviews. The team leaders, having spent much time with mathematics or science education over long careers, looked at these reform projects through several lenses; the teams sifted through the mountains of data they had collected in order to tell the story of each project in rich detail. We encourage the reader to peruse Volume 2 (the five science case studies) and Volume 3 (the three mathematics case studies) of *Bold Ventures* to take full advantage of the research effort invested in studying these landmark examples of science and mathematics education reform. The present volume represents a further distillation of this research—a synthesis based on the case studies themselves and on a series of exciting, if exhausting, meetings of all team members two or three times a year designed to delineate and clarify the cross-case themes as they emerged from the developing work.

The U.S. research was part of a larger international effort to study innovations in science, mathematics, and technology education undertaken by 13 member countries of the Organisation for Economic Co-operation and Development (OECD). The genesis of the international case study project was the countries' shared dissatisfaction with the state of education in these subjects; the participating countries wished to be informed about improvement efforts and their impact elsewhere in the industrialized world. The result was the largest qualitative research project ever undertaken across countries. The work has aroused much interest. In addition to the publication of a monograph based on the case studies from all the countries, two internationally sponsored conferences—one in the United States and one in Europe—have been held to disseminate key findings. In addition, several OECD countries organized their own dissemination conferences.

Just as we in the United States are learning from reform efforts in other countries, the stories and findings of the U.S. cases are proving instructive to them. Innovation in our large, decentralized, and diverse country is of particular interest for education ministries in formerly centralized systems that are now experimenting with decentralized curriculum reform and greater autonomy for teachers. Educators in other countries can compare policy priorities in the United States to issues considered important in their own countries. Also, documentation of the U.S. reform efforts in the three volumes of the *Bold Ventures* series will complement and help in interpreting—both for this country and for an international audience—the quantitative results generated on student achievement, current curriculum, and teacher and student background by the Third International Mathematics and Science Study.

Audiences

Bold Ventures is intended primarily for people working to improve science and mathematics education, though we believe it holds much of interest for audiences concerned with educational reform more generally. We include among these audiences:

- policymakers in a position to influence schools and educational systems;

- teacher educators and staff developers working with prospective or already practicing teachers;

- science, mathematics, and engineering professionals;

- school administrators;

- teachers of science and mathematics; and

- especially, those who would join the innovators and reformers of science and mathematics education.

The three volumes in the series unravel the origins, development, and implementation stories of eight of this country's major reform initiatives. They address questions surrounding the "what" as well as the "how" of these innovations. Curriculum specialists, researchers, and teachers of science and mathematics will find the detailed case studies in the other two volumes of *Bold Ventures* of particular interest and practical utility in the teaching and learning of science and mathematics. Policymakers, administrators, and those interested in the progress and process of educational reform should find this cross-case volume informative and helpful in understanding the origins and impacts of innovation in these two critically important school subjects.

Organization

While the authors of each individual chapter bear primary responsibility for its content and language, the book as a whole is a collaborative effort of the eight research teams and the editors. The book starts with brief summaries of the OECD case study project and the U.S. innovations studied and an overview, with two examples, of the methodology used by the U.S. research teams. Chapter 2 discusses the motivation for reform of mathematics and science education in the United States and the forces shaping reform initiatives in the last decade, including prominent actors and strategies, as well as the micro forces shaping individual projects. Chapters 3, 4, and 5 elaborate the main themes drawn from the eight case studies: Mike Atkin, Jeremy Kilpatrick, and their coauthors discuss changes in the fields of mathematics and science and concomitant changes in the corresponding school subjects; Norman Webb addresses the changing roles of teachers; and Senta Raizen, Douglas McLeod, and Mary Budd Rowe review the changing conceptions of educational reform as played out in the innovations. In chapter 6, Robert Stake and Senta Raizen deal with some issues we found—somewhat surprisingly—underemphasized in the U.S. case studies, including assessment of student learning, external evaluation of the projects, and equity concerns. Chapter 7, authored by Michael Huberman, takes a close look at the implementation of the eight innovations in light of their ongoing development. The book ends with some concluding thoughts about both the reform projects and our study of them. In the appendix, we provide brief summaries of the 15 case studies conducted by the other participating OECD countries.

Acknowledgments

We gratefully acknowledge the many individuals who contributed to the research effort reported in this and the two companion volumes. Just counting the researchers, advisors, and support staff, over 40 U.S. professionals invested substantial amounts of their time in this effort over the last four or more years. Most of these individuals are cited on other pages of this volume: advisors are listed just before the contents page, and short biographies of senior authors are included in the section on contributors. We thank the indispensable administrative assistants to this project, especially those who spent large amounts of time organizing meetings, preparing briefing books, and working with manuscripts: Susan Callan and LaDonna Dickerson at the National Center for Improving Science Education, Sally Lesher at the University of Wisconsin, and Sunny Toy at Stanford University. Nita Congress and her colleagues toiled to help us line edit this rather large body of work; Shelley Wetzel of Marketing Options handled the manuscript layout.

The Department of Education and the National Science Foundation (NSF) have supported this work on an equal basis, as administered by NSF under grant number RED-9255247. We are grateful for their funding and supportive monitoring. The content of this report does not necessarily reflect the views of the Department or NSF. Their funding extended beyond the research to support dissemination efforts including a 1993 workshop to acquaint potential audiences with the goals of the U.S. case studies; and a 1996 conference in Washington, "Getting the Word Out," to release the U.S. and international results. We also gratefully acknowledge the support of the Mellon Foundation for the editing and production of this book.

Eve Bither, Director of the National Educational Research Policy and Priorities Board, spearheaded the Department of Education's support of this project and has steadfastly advanced our work. Several current and former officers at NSF have generously assisted us over the years: Ken Travers, now of the University of Illinois; Daryl Chubin, director of Research, Evaluation, and Communication; David Jenness, now an independent consultant; Conrad Katzenmeyer, senior program officer; Iris Rotberg, program officer; and Larry Suter, deputy division director.

In closing, we wish to applaud and thank those who have made our work possible—the reform-minded professionals who worked in and out of the classroom to make their visions of better science and mathematics education come to life. Many people were gracious enough to let us witness their endeavors again and again and to speak with them at length. When busy educators and innovators welcome researchers into their world, it behooves these researchers to take the greatest care in depicting their efforts. We and our colleagues hope the many individuals who generously gave of themselves and their time will find *Bold Ventures* to be respectful, accurate, and helpful in advancing their important work.

Senta A. Raizen
Edward D. Britton

Washington, D.C.
November 1996

1

Study Background

Senta A. Raizen

This book presents a distillation of the major themes emerging from case studies, carried out over three years, of eight U.S. reform initiatives in science and mathematics education. In parallel, 12 other member countries of the Organisation for Economic Co-operation and Development (OECD) conducted case studies of innovations in science, mathematics, and technology education.[1] The participating countries were Australia, Austria, Canada, France, Germany, Ireland, Japan, the Netherlands, Norway, Scotland, Spain, and Switzerland. A brief description of these countries' studies is given in the appendix. When appropriate, we draw from these case studies to enlarge on the U.S. findings.

The idea for conducting international case studies arose from a growing concern, shared by OECD member nations, for more effective programs in mathematics, science, and technology education. Moreover, the participating countries recognized the need to have a better understanding of the policies, programs, and practices that lead to successful outcomes in these fields. What is the nature of the changes being formulated? How are such policies, programs, and practices developed? How are they implemented in settings where they are successful? What processes support implementation? What decisive roles are played? What outcomes are attained? The desire to have answers to these questions led the 13 countries to carry out one or more case studies, using common guidelines.[2]

But why should questions about science and mathematics education be asked in an international context? As the United States and other countries become more globally oriented and interdependent—not only in terms of economic activity, but also in terms of the environment, movement across national boundaries, educational and cultural exchanges, and even day-to-day existence—they are discovering many common concerns. In education, in particular, policymakers and practitioners feel they need to look beyond their own shores to

---

[1]Three of these countries studied the introduction into precollegiate education of technology as a school subject. The U.S. work included no such case, though three of the cases did involve the use of educational technology in science or mathematics.

[2]Countries carrying out two case studies were Canada, Japan, and Norway.

bring the best available knowledge to bear on the improvement of traditional practices. Explorations by several OECD countries of current reform efforts in science and mathematics education had demonstrated a number of parallel developments, which could be augmented and reinforced by sharing knowledge and experience. Therefore, the case study project was undertaken to highlight noteworthy developments in science, mathematics, and technology education across the participating countries. More important still was the perceived need to provide documentation that would expand understanding of how educational innovations occur and are implemented in different settings. Thus, an underlying assumption of the OECD case studies project has been that the context, characteristics, and implementation features of significant innovation within a given country will turn out to be informative and productive for science and mathematics education elsewhere.

The International Project

The origin of the international Science, Mathematics and Technology Education project dates back to 1989, when a meeting of interested researchers resulted in an issues paper, further refined during a 1990 meeting of experts from 18 OECD member countries, including the United States. The group defined the goal of the project as providing models and other information to help member countries improve curriculum and instruction in science, mathematics, and technology education. They also agreed on the main themes for the project that each case study should seek to address:

- The context (historic, social, political, educational) within which the innovation was formulated;

- Processes by which change was implemented, both as envisioned in planning and as actually experienced;

- Goals and content of the innovation;

- Perspectives of students;

- Settings for learning, instruction, and evaluation;

- Methods, materials, and equipment; and

- Teachers, teacher education and training, and other forms of support.

In terms of the three subject areas, researchers were to address:

- Redefinitions of the scope and structure of science and mathematics content;

- The role of problem-solving approaches and other applications in building mathematical and science knowledge and skills;

- The extension of science and mathematics education to a more diverse student population;

- Ways in which technology education is implemented in the general curriculum of elementary and secondary schools; and

- The interrelationships among science, mathematics, and technology in the school curriculum.

The outcome of the international project was *Changing the Subject: Innovations in Science, Mathematics and Technology Education* (Black and Atkin 1996). This book distills the major findings from all 23 cases, concentrating on the themes of greatest concern to the participating countries. These international findings are summarized in *A Global Revolution in Science, Mathematics, and Technology Education* (Atkin et al. 1996a). These insights complement those of the Third International Mathematics and Science Study (TIMSS), which has collected data on a number of contextual factors as well as on student achievement in science and mathematics in 40 to 50 countries. However, TIMSS was not able to examine in detail how these factors operate at specific sites.

The international case study project supplies updated information to American educators and policymakers on reform efforts and their implementation in other industrialized countries. The findings from the case studies provided in the OECD book help illuminate how policy, organizational, and classroom factors contribute to the success or failure of a given innovation in a specific country. Consideration of the goals and progress of reform in other countries can help the United States in developing programs to achieve its own policy goals in science and mathematics education. The OECD project also has created an international network of educators and researchers by fostering continued interpersonal contacts and exchanges of information.

The U.S. Studies

The development of the U.S. case studies occurred in two phases, both funded by the U.S. Department of Education and the National Science Foundation (NSF). In the first phase, the National Center for Improving Science Education (NCISE) and the National Center for Research in Mathematical Sciences Education selected eight innovations for case study development and prepared 20-page descriptions of each. These descriptions were published by OECD

(1993) under the title *Science and Mathematics Education in the United States: Eight Innovations*. The second phase consisted of the in-depth study of each of the innovations and the cross-case analyses summarized in this book.[3]

In selecting innovations for intensive study, we aimed for a varied and balanced set with respect to subject areas, grade levels, the nature of the interventions demanded by the innovation, and the origins and main driving forces behind the innovation. Table 1–1 provides an overview of the range of the selected innovations.

Table 1-1.   Characteristics of U.S. Innovations Studied

	Curriculum Development		Comprehensive		
	K-8	9-12	District	State	National
Mathematics		Contemporary Precalculus*	Urban Mathematics Collaborative		NCTM Standards
Science	Kids Network* Mimi*	ChemCom		California Science Education Reform	Project 2061

*While the primary focus of these innovations was new approaches to science or mathematics, they also developed new uses of technology in instruction.

Three of the cases are in mathematics education: the Contemporary Precalculus course, the National Council of Teachers of Mathematics (NCTM) Standards, and the Urban Mathematics Collaborative (UMC) project. Five are in science education: the California Science Education Reform initiatives, the Chemistry in the Community (ChemCom) course, the Kids Network units, Project 2061, and the Voyage of the Mimi materials.[4] Two of the innovations (Kids Network and Mimi) use communications technology and multimedia approaches for instruction. Four of the projects are focused on innovative curriculum materials; four are more comprehensive. Two of the materials develop-

---

[3]Seven of the innovations studied in depth are the same as those summarized in the OECD 1993 report; for the eighth innovation, the Precalculus project was substituted for the California Restructuring of Mathematics Education (Webb 1993). The reason for the substitution was that the Precalculus project represents a unique grassroots effort by teachers, whereas a state initiative already was represented in the California Science Education Reform case study.

[4]The Voyage of the Mimi was designed to be used as a tool in both science and mathematics instruction. In practice, however, teachers have emphasized science much more than mathematics.

ment projects are aimed at the high school level (ChemCom, Precalculus); one is intended for the middle school and upper elementary grades (Mimi) and one is designed for the upper elementary level (Kids Network).[5]

The origins of the innovations also are varied. Two are large national initiatives (NCTM Standards, Project 2061); one was developed through an NSF grants competition (Kids Network); two came about through initiatives within a professional school or organization and subsequently received federal and foundation funding (Mimi and ChemCom). One innovation represents a state initiative (California reform); one was conceived by a private foundation to focus on large urban school systems (UMC); and one came from the initiative of teachers at a public school emphasizing science and mathematics (Precalculus).

It is worth noting here that, with the exception of the German and Swiss projects, all the international innovations comprising the OECD project were sponsored directly by the government or undertaken in response to a governmental mandate. In contrast, only one of the U.S. innovations studied—the California reform—was initiated by a government agency, although several of the other projects received funding from federal agencies (principally NSF and the Department of Education) during some portion of their work. The selection of innovations for study also was a governmental decision for all countries but the United States.[6] Here, the selection was made by the case study researchers, based on advice from a representative group of experts, i.e., the United States project's Advisory Board, convened by the National Center for Improving Science Education.[7]

Recent data confirm that cases discussed in this volume are well-known across the country and, therefore, very relevant selections for a national study of major innovations in science education. In their 1993 survey, Weiss et al. (1994) found that 67 percent of elementary and middle school teachers had heard of the Kids Network; 49 percent of teachers for grades 5–8 knew of the Second Voyage of the Mimi; and 58 percent of all high school science teachers (not just chemistry teachers) knew Chemistry in the Community. Ten percent of teachers at every type of school (elementary, middle, and high) reported making "considerable use" of Project 2061's Science for All Americans—a strong influence for something that is not a textbook nor a specific pedagogical technique. Weiss et al. collected considerable data on teachers' familiarity with the NCTM standards. In Volume 3 of *Bold Ventures*, McLeod et al. (1996, pp. 86–87) review the survey findings; they summarize that most teachers were informed about the standards, and many were prepared to explain them to their colleagues.

---

[5]At the time of the study, National Geographic only marketed an elementary school version of Kids Network. Now, there are units designed for middle school as well.

[6]This may, at least in part, be due to the nature of OECD, since member countries are represented by government officials.

[7]Members of the U.S. project's Advisory Board are listed in the front of this volume.

In what ways does each of the selected innovations represent a significant change? How notable are these changes for U.S. education? Why is the innovation worth studying? What can be learned from each case? These and many other questions find responses in the documented narratives of the eight innovations. The synthesis in this book of major themes drawn from the eight U.S. cases in no way adequately reflects the detail needed to understand the cases. The case stories are published in two companion volumes in the *Bold Ventures Series*—*Volume 2: Case Studies of U.S. Innovations in Science Education* and *Volume 3: Case Studies of U.S. Innovations in Mathematics Education.* We encourage the reader to take up these two volumes for the immersion the cases deserve.[8]   However, to provide the necessary background for the rest of this book, we briefly summarize the innovations and the case studies of them. (The heading used for each innovation is the title of the respective case study as published in Volumes 2 and 3 of *Bold Ventures*.)

Building on Strength: Changing Science Teaching in California Public Schools. California has led the way in systemically influencing science instruction through concerted, congruent actions in four policy arenas: (1) reconceptualizing the science curriculum in the innovative *Science Framework for California Public Schools* (California Department of Education 1990), (2) developing new statewide assessments to support the desired changes in science instruction, (3) creating means for helping teachers change their science instruction, and (4) requiring instructional materials to be consistent with the framework. The case study analyzes the changes in approach to science content and teaching required by the reform and describes in detail California's reliance on teacher networks to implement the desired changes.

ChemCom's Evolution: Development, Spread, and Adaptation. This course, popularly known as ChemCom, is an alternative to traditional high school chemistry. The curriculum focuses on chemistry as it is needed to understand sociotechnological problems rather than on chemistry as organized along traditional disciplinary lines. Through this approach, the course is designed to attract nontechnical, college-bound students, but it also is being taken by students not planning to attend college. In quantitative terms, ChemCom is widely taught here and abroad. ChemCom development was supported in part by NSF, then distributed through a commercial textbook publisher of curriculum materials. The case study details the development of the course; its implementation; and the role played by the American Chemical Society, the sponsoring society.

The Challenges of Bringing the Kids Network to the Classroom. Through telecommunications, the Kids Network enables upper elementary school students across the country—and in other countries—to exchange scien-

[8]Both volumes are available from the National Center for Improving Science Education (Raizen and Britton 1996a and 1996b). The senior authors of these works are J. Myron Atkin, Michael Huberman, and Mary Budd Rowe (Volume 2); and Jeremy Kilpatrick, Douglas B. McLeod, Robert E. Stake, and Norman L. Webb (Volume 3).

tific data they collect during investigations of real-world environmental problems—for example, acid rain, waste disposal, and water pollution—and join with scientists to analyze results. The Kids Network was developed by the Technical Education Research Centers, principally through support from NSF, and is distributed by the National Geographic Society. This innovation, as is the case for Voyage of the Mimi (described below), is of interest not only for its creative incorporation of technology, but also for the story of its implementation. The case study describes how the project provides a medium of change for teachers and classrooms, how teachers implement the curriculum, the experience of students, and incentives and barriers related to the use of the technology.

Teaching and Learning Cross-Country Mathematics: A Story of Innovation in Precalculus. This grassroots effort differs markedly from most curriculum projects. A group of teachers at a public high shool in North Carolina emphasizing science and mathematics developed a precalculus course based on the modeling of real phenomena. It emphasizes applications, data analysis, and matrices; the use of graphing calculators also is integral to the course. Through word of mouth, teachers in other locales learned of the course and began to use it. This in turn enabled the North Carolina teachers to obtain NSF funding for refining the materials and also to secure a commercial publisher. The case study describes how the original group of teachers became involved in developing the course, holding workshops for other teachers interested in changing their mathematics instruction, and thus spreading their approach to high schools across the country.

The Different Worlds of Project 2061. This initiative, led by the American Association for the Advancement of Science and supported initially by several foundations, endeavors to fundamentally restructure science, mathematics, and technology education across all precollegiate schooling. The project, launched in 1985 when Halley's Comet was in earth's vicinity, is named for the year of the comet's next return, to indicate the project's long-range nature. The project's first major publication, *Science for All Americans* (AAAS 1989), sets out a vision for what science is important for all to know. A second publication has provided more detail in the form of "benchmarks" on what students should know and be able to do at various grade levels. Project 2061 intends to translate these documents into new school contexts for science, new curricular programs and materials, and new instructional practices. Six school district centers are working with the project to develop curricular approaches and apply the benchmarks. The case study examines the development of the publications, their degree of national impact, and their relationship to the curriculum development in the six school districts.

Setting the Standards: NCTM's Role in the Reform of Mathematics Education. In 1989, NCTM published *Curriculum and Evaluation Standards for School Mathematics*; in 1991, there followed *Professional Standards for Teaching Mathematics*; and in 1995, *Assessment Standards for School Mathematics*. NCTM is promoting a systematic reform of mathematics education, emphasizing a new conception of mathematics as problem solving, reasoning, communication, and connection. NCTM is the first subject-area association to create a comprehensive series of standards documents. Further, NCTM did this on its own initiative, over a long period of time, and without government resources. The case study tells the story of how a professional organization of mathematics teachers came to assume national leadership in the field of mathematics education and thus to influence national and state policy in the development of high educational standards. The study also looks at how the standards are being interpreted in schools.

The Urban Mathematics Collaborative Project: A Study of Teacher, Community, and Reform. The UMC project is aimed at improving mathematics education in inner-city schools, while at the same time identifying new models for meeting the ongoing professional needs of teachers. Funded by the Ford Foundation, mathematics teachers in 16 school districts and others in their communities have worked together with the objective of improving the quality of mathematics teaching in the districts' middle schools. The case study reports on the interaction among business, industry, higher education, school systems, and teachers in their quest to create change in the teaching of mathematics. It chronicles the career trajectories of teachers as they assumed leadership roles, as well as describing how new conceptions and practices emerged in the mathematics classroom. It analyzes the factors that facilitated—and those that impeded—institutionalization of this approach; for example, the problems associated with using a volunteer model.

Science, Technology, and Story: Implementing the Voyage of the Mimi. The Voyage of the Mimi is a curriculum program for 9- to 14-year-olds that combines videos or videodiscs, computer software, and print materials to present an integrated set of concepts in mathematics, science, social studies, and language arts. The project builds on a storyline of expeditions involving a research ship called the Mimi. The first *Voyage of the Mimi* takes students on a study of whales off the coast of New England. In addition to studying whales, students apply mathematical ideas of proportional reasoning and triangulation to navigation problems. In *The Second Voyage of the Mimi*, students go on an archeological expedition in the Yucatan Peninsula of Mexico where they work with the Mayan number system and Mayan calendar and study the relationship between the earth and sun. The case study relates how the course units were introduced into schools, how teachers collaborated to teach them, and how they adapted these materials. It describes teachers' reactions to the materials and analyzes factors that make for or impede effective implementation.

Methodology

We chose the case study approach as our main methodology for studying the innovations. This is the method of choice for a close-up, in-depth understanding of their origins and implementation. Case studies also allow researchers to deal with latent or underlying issues that might elude the survey analyst or the short-term field observer. For example, an understanding of site dynamics can evolve only gradually, from the time when a project is just under way to the time when it has spread widely enough to allow the formulation and testing of hypotheses about what facilitates or inhibits successful implementation. All the same, case study research admits that there can be several, varying interpretations, both among actors at the research sites and between these actors and the field researchers. We have tried to reflect these differences in interpretation in our individual case reports (see Volumes 2 and 3 of this series). The case study researchers attempted to reconstruct critical events that occurred before they picked up the case; they were then in the field for a year (in some cases, longer), following project activities and developments as they unfolded at selected sites. Nevertheless, each of the case reports records a "snapshot in time"—the situation and context as they were observed and recorded during a specific period in the life of each project.

The Research Questions

Each of the case studies addressed some general research questions; a second group of questions was particularly pertinent to the more comprehensive efforts; a third group consisted of more detailed questions applicable to implementation of curricular materials. The three groups of questions are given below. Specific questions addressed by each case study are elaborated in the individual reports.

General Questions.

- What motivated the genesis of the innovation? Who were the key movers? What were the contexts of development—social, historical, institutional, political?

- What was the problem or dilemma addressed by the innovation?

- What were the content, design, and underlying assumptions of the innovation? How much variation in point of view was there in different settings?

- How was progress or success defined, assessed, and known? Which criteria were used by different key actors or institutions?

- What were the principal facilitators and constraints in the design and execution of the innovation?

- What was the involvement of teachers in the different U.S. innovations? How did teacher involvement affect the introduction and acceptance of innovations?

Additional Questions for Comprehensive Efforts. This next set of questions is particularly appropriate for such comprehensive efforts as the UMC, NCTM Standards, Project 2061, and the California Science Education reform.

- What policies were advocated? What was the rationale for these policies? Which of the policies were, in fact, instituted?

- For state- or district-focused efforts, what direct funds or indirect resources were allocated to the reform?

- Did the effort generate a favorable climate for science or mathematics education reform—nationally, in states, in districts?

- What roles were envisioned for business and industry, for institutions of higher education, and for other nonscholastic organizations?

- Did collaboration occur among professionals working in different contexts (scientists or mathematicians in higher education or industry, teachers, school administrators)? If so, in what forms?

- Did local users modify the intended innovations, and, if so, for what reasons?

Additional Implementation Questions. These questions are addressed by the four case studies concerned with the development of curriculum materials for science or mathematics and their implementation: ChemCom, the Kids Network, Precalculus, and Voyage of the Mimi. Except for ChemCom, these innovations involved issues of using computers and other technical devices as an integral part of the instruction.

- What were the key marker events of each phase of implementation, and how did these events affect the outcomes of local implementation?

- What, if any, resources and adjustments in working conditions did schools or districts provide to facilitate use of project materials?

- What did the innovation look like on a daily basis? What was the correspondence between teachers' and students' perceptions of the innovation in action?

- What, if any, changes did teachers make in their instructional practices during successive uses of the materials? What, if any, training and other supports were available?

- Were schools and classrooms equipped technologically and with needed supplies to carry out the new curriculum or the innovation's instructional and learning strategies?

The Case Study Research Cycle

The overall work of studying the eight U.S. innovations took three years, as shown in table 1–2. The first year (1992-93) was devoted to three tasks. The first task was to identify, within the OECD framework, the common research questions covering all the cases, as well as project-specific ones. The second task was to identify potential sites for in-depth study. Quite different approaches were needed for the individual cases, ranging from developing a sampling plan for innovations already in wide use (as was the case for several of the curriculum materials projects) to selecting from a few extant sites (for example, for Project 2061 or the UMC project). The third task consisted of developing specific methodologies: identifying key project informants to be interviewed, formulating appropriate surveys and questionnaires, and creating interview and observation protocols. (Each of the case study reports in Volumes 2 and 3 provides further methodological details germane to the individual case.)

The second year (1993-94) was devoted to intensive field work; development of analytical methods, including coding of observations and interviews; analysis of critical documents; preliminary analyses of field data to generate further field work; and interaction with potential audiences to generate dissemination plans for the study results. Intermediate analyses, reviews, and feedback from the projects and sites as well as from outside consultants; final analyses; and writing of each case and this cross-case volume consumed the third year (1994-95).

Two Examples of Research Methodology

The brief descriptions given below of the methodologies used in two of the case studies—one curriculum project, one more comprehensive project—illustrate some of the differences and similarities in research approaches.

The Mimi Case Study. The case study research team looked closely at six exemplars of Mimi during 1993-94. Collectively, researchers and informants (teachers, administrators, students) reconstituted the local genesis of Mimi and the story of its implementation, documented the constraints and supports along the way, and recorded its daily enactment within the classroom. Visits by the research team allowed local participants to describe their perceptions of Mimi and tell their personal history of involvement with its materials and activities. The research team spent much of its time observing how Mimi was acted out: how the teachers' guides were actually used, how teachers modified or otherwise experimented with the curriculum, and, above all, what pupils and teachers actually made of program activities.

Site Selection. Criteria used to select study sites included: (1) location of the school (urban, suburban, rural); (2) socioeconomic status and racial mix of student population; and (3) variations in local practice. In addition, researchers selected for grade level (elementary or middle schools), use of Mimi for three or more years, self-contained classrooms or teaching within a subject

Table 1-2.   Research Cycle for Eight Individual Cases

1989	OECD meeting: design of international study, including research themes.
1991-92	U.S. pilot study: researchers identify and describe potential U.S. cases; Advisory Board selects eight innovations for extensive case studies; brief descriptions of U.S. innovations published by OECD (1993).
1992-93	Refinement of research questions, analysis of documents, development of instruments, sampling of sites, start of field work.
1993-94	Main data collection and analysis: observation of classrooms and key events, interviews of key individuals—initiators, developers, and teachers. U.S. Department of Education and NSF sponsor an NCISE "dissemination" workshop to alert potential audiences to the research and elicit suggestions for directions of special interest and importance.
1994-95	Completion of data analysis, drafting of reports.
1995-96	Editing and publication of *Bold Ventures* in three volumes. Release of synopsis of results in *Education Week*'s "Forum" series, April 10, 1996; ED, NSF, OECD, and NCISE conduct "Getting the Word Out," a conference in April 1996 to highlight international and U.S. results.

area (for example, science), and collaborations with other users within the school or district. Visits to potential sites were key to making the final selection. All six sites ultimately selected were in the Northeast. Although this was done for reasons of convenience, the researchers ascertained that the variety within this region was as great as could be found in the rest of the country. The six sites were located in five states: There were two in rural areas, two in suburban settings, and two in urban settings; three were elementary schools, and three were middle schools.

Instrumentation. Several versions of the interview and observation schedules were piloted. The final set of instruments included the following:

- interview guides for developers, school- and district-level administrators, teachers, pupils, and assistance providers;

- observation schedules and checklists for classroom activity;

- guidelines for classroom observation; and

- formats for analysis of documents, work samples from the site, analytic memos, and site visit reports.

Data Collection Procedures. In the 1993–94 school year, the six schools were visited at least three times, one to three days at a time, with a total of 45 days on site. Site visits, planned in cooperation with sites, typically began with the start of the school day and ended with school closing. The researcher observed lessons and activities related directly to the Voyage of the Mimi, and compared them with lessons in mathematics, social studies, and reading. Two to six observations of Mimi-related activities, one-half to two hours in length, were made for each participating teacher. In addition, formal and informal interviews were conducted with teachers, ranging in length from one-half to three hours. Principals, administrators, science specialists, and students also were interviewed.

In summary, more than 100 hours of formal interviews with 17 teachers, 66 students, and 11 principals and other support staff were carried out, and an estimated 135 hours of observations were conducted. In addition, the researchers joined classes from three sites for their all-day field trips: an ecology camp, a whale watch, and a regional "Mimi Festival."

Research Questions. Research questions were organized into the following categories:

- Origins—the history of program development and the initial characteristics of the program;

- Process of local adoption—the circumstances by which the innovation was adopted at the site and the local context at the time of adoption;

- Institutional issues—the "fit" of the innovation to the organization, including the innovation's characteristics (philosophy of mathematics and science, use of technology, multicultural sensitivity);

- Roles of principal and district office personnel—their knowledge of the innovation, their technical and social support, provision of assistance;

- Implementation of the innovation—project influence on core aspects of teachers' work and pupils' understandings and perceptions;

- Pupil perceptions—engagement with and reactions to the typical activities, the materials in use, and the tasks undertaken; and

- Stabilization of the innovation—effects of the school organization and yearly routines on the operations and substance of Mimi.

Analytic Procedures. Data analysis was largely iterative; i.e., it proceeded as the data were collected and reoriented further data-gathering and interim analysis. Eventually, 17 functional coding categories were created from 32 thematic codes and 173 general codes. The thematic material, keyed to the research questions, was then subjected to a question-by-question analysis using templates devised by Miles and Huberman (1994). In particular, checklist matrices, event listings, and time- and role-order matrices were used to gather similarly plotted data on one chart. The same was done with conceptually clustered matrices and causal chains. Comparative analysis then allowed for clustering cases, establishing significant similarities and differences across cases, and, from there, identifying systematic sources of variation. This provided the explanatory material, as well as the narrative, to account for Mimi across and within the six settings—for its history, evolution, and instructional activities over time.

The Case Study of the Urban Mathematics Collaborative Project. At the start of the case study in 1993, coordinators and directors of all 16 UMC sites were mailed a letter describing the OECD case studies research project and informing them of a proposed telephone interview of about 45 minutes. An individual at each of the 16 collaborative sites participated in this interview. The interviews consisted of questions designed to establish the changes that had taken place in each collaborative since 1990 when they had last been documented, the factors that contributed to such changes, and the current state of administration and activities.

Site Selection. Following the contact interviews, five collaboratives were chosen for further exploration as representative of the 16 collaboratives: Columbus, Los Angeles, Memphis, Milwaukee, and Philadelphia. These five sites offered a cross section of the UMC project in several ways: (1) Geographically, the five sites were spread across the United States, with most major population regions represented. (2) Several of the best models of collaborative activity were found among the five selected sites. (3) The Los Angeles collaborative served the nation's second largest school district, the Columbus collaborative served a small urban school district and numerous very small rural school districts, and the other three sites served school districts that fell between these two extremes. The chief shortcoming of the site selection was the failure to select one of the two sites that had not survived so that reasons for their failure to thrive could have been established.

Site Visits. A two-day introductory site visit to each collaborative was conducted during fall and winter of 1993-94 to determine the most significant themes to be explored. The researchers interviewed staff and administrators, teachers, and participants from business, higher education, and school and district administration. Principal questions asked were:

- What were the identifiable stages of growth of the collaborative over its existence?

- What strategies did the collaborative employ to enact change?

- Who were the key players in the collaborative in instigating and carrying out change?

- What changes in teachers' roles in or out of the classroom occurred in connection with the collaborative?

During the spring of 1994, a major site visit of three to five days was made to each of the five collaboratives. The main purposes of the major site visits were to collect data that would inform the themes emerging in the ongoing analysis and to immerse the researchers in the context of the collaborative sites. The visits included interviews with teachers and other participants from all the sectors, as well as observations of mathematics classrooms and collaborative activities. Additional information was gathered through informal conversations and interactions.

Principal questions pursued were:

- How did mathematics education in the classroom (content, instruction, assessment, teacher and student roles) change in relation to the collaborative?

- What educational goals did teachers and others hold for students with regard to mathematics? How were these goals related to the collaborative? How were these goals manifested?

- What professional goals did teachers hold for themselves or did others hold for teachers? How were these goals related to the collaborative? How were these goals manifested?

Principal areas of classroom observation were:

- objectives, goals, and purposes of mathematics lessons;

- discourse in the classroom between teachers and students and among students;

- engagement of students with challenging mathematical content; and

- activities and roles of teachers and students in mathematics learning.

Brief exit interviews were conducted via telephone during the spring of 1995 with one or two administrators or teacher-leaders at each of the five selected collaborative sites. The main purpose of the exit interviews was to obtain, update, and verify specific information regarding the constituencies of the collaboratives, the specific goals of each collaborative, and the strategies used to achieve those goals.

Analysis. During the data collection periods, the researchers who conducted each site visit offered reports, impressions, and insights regarding the visit at subsequent research team meetings. The team met two to four times each month to articulate themes and areas of interest—generally focused on individual sites—as well as to handle technical details of the case study.

The text of all interviews was electronically processed using Hyper-RESEARCH (Hesse-Bieber et al. 1993), a computer application designed to facilitate qualitative research. The principal purpose of the coding was to extract evidence from the data relevant to each area of interest and theme. The coding scheme used consisted of 4 primary codes, 14 secondary codes, and between 63 and 66 tertiary codes. A series of matrices (site × theme) was generated to aggregate information within and across sites. As the matrices evolved, they were used to generate and modify the case study's coding scheme and to generate outlines for the writing phase of the project. Validation of themes and areas of interest focused on both the individual sites and the entire case during periods of intensive analysis.

A major concern for the researchers was the need to establish connections between the goals and activities of the collaboratives and changes that were described by participants in the collaboratives or observed by the researchers during data collection. The researchers sought corroborating evidence of change from several sources, including interview data from individuals with different roles in the collaboratives' operations, observation data, and printed material, as appropriate. The links between change and the collaboratives—which the researchers conjectured—generally arose from impressions and insights following data collection. The coding of interview data and examination of observation notes and printed material provided potential evidence to substantiate the researchers' conjectures. Importantly, the discussions and deliberations of the research team also served as a forum for testing the validity of conjectures about change in relation to collaboratives. The combined perspective of the research team was viewed as a critical filter for eliminating the biases and preconceptions of each individual researcher.

Writing was a vital component of the analysis and construction of the UMC case study. During the three years of the study, several hundred pages were written in the form of reports, memos, and draft documents. The writing process served as a vehicle to record questions, insights, impressions, and results of interim analysis. The nature, strength, or lack of evidence for change and for claims only became fully apparent during the writing process. The final report of the case study represents what the researchers consider to be the most important contributions of the UMC project as a living example of innovation in mathematics education.

Summary Comments on Methodology[9]

The methodology used in this project was noteworthy, and possibly unique. Multiple-case studies are, in themselves, difficult undertakings. In this instance, the choice of eight distinct projects made the enterprise harder. There was no sampling by type of project, scale of project, presumed level of outcome, length

---

[9]These comments were contributed by Michael Huberman in reflecting—in his dual role as research team leader and expert methodologist—on the three-year process of the entire case study project.

of life, or level of core activity. Cross-case findings, overarching themes, leit-motifs, and common answers to research questions could only emerge when findings were strikingly similar or contrasted. Those cross-case findings are reported in this volume.

What is especially noteworthy was the process of research. The adjectives that come to mind are "interactive," "collective," "iterative." As described in this volume, the research questions and sample of projects were derived by a panel of experts, who were familiar both with previous research in science and mathematics education and with ongoing reforms. With the creation of study teams in disparate areas of the country, each with a subset of projects to study, the issue of coordination came to the fore. For virtually all participants, the option to manage the study collectively seemed an inspired one. At regular intervals, the study teams—including research associates—would come together and work through the instrumentation, sampling, data collection and, progressively, questions of analysis and interpretation. The role of NCISE coordination was to pool the results of these interim meetings, structure the next steps, and organize the following session.

In this way, findings from each study were fed into a collective forum as they emerged. They could then be compared with other substudies, discussed with other teams, and probed for emerging commonalities and distinctions. At the same time, collective scrutiny was a strong—and not always painless—stimulant to heightening reliability, guarding against bias, making a stronger case for emerging conclusions, and trying on leitmotifs or emerging trends to see if they really fit the data set. Typically, researchers went back home overstimulated, but also with an agenda for testing findings more stringently or for pursuing leads that had not occurred to them before.

At the end—much as in a funnel—meetings addressed only the major findings, the most salient variables, the most compelling explanations. This gave shape and focus to the final writeups, while still allowing the descriptive and inferential latitude that case histories require.

There are no grounds for romanticizing the process. It did, however, unequivocally enhance both the reliability and the validity of the project as a whole. Its secondary effects are also noteworthy. Seldom have senior researchers interacted as intensively and extensively on a single project. And seldom have junior researchers been included as equal participants in presentations, comparisons, "buzz" sessions to create new frameworks or interpretive grids, and confrontations on method and substance—all relative to settings they virtually lived in and data of which they had the most first-hand knowledge. This, as one senior researcher put it, was a moveable intellectual feast. It will be judged on the merits of its contribution.

In sum, the research effort was characterized by close collaboration among the research teams. Teams met collectively several times each year, first to decide on the common research questions and general approaches to the field work and site selection; and later to exchange early findings, develop common themes for the cross-case analysis, and decide on the length and style of the case reports. Still later, project meetings discussed key chapters of the cross-case volume and addressed dissemination issues. Several of the meetings also involved reviews and critiques by Advisory Board members. Several indicators suggest that this unusually dense collaboration among the eight research teams proved motivating and stimulating to all involved, but especially to the graduate students serving as research associates in the project. This mode of working may also have enriched the quality of both the individual case studies and the cross-case volume.

2

The General Context for Reform

Senta A. Raizen

In this chapter, we provide the backdrop for the eight innovations we studied—what gave rise to the drive for improving education in general and science and mathematics education in grades K-12 in particular. We note some general reform directions, how they affected the eight innovations, and, in turn, how they were affected by them. The major themes that developed from our research, referred to briefly in this chapter, are discussed in greater detail in succeeding chapters.

Reform of elementary and secondary education has been a hallmark of U.S. education policy of the last decade. The current reform efforts were sparked by a number of reports published in the early 1980s decrying the state of the nation's schools—prominent among them the report *A Nation at Risk* by the National Commission on Excellence in Education (1983). Several of these reports specifically documented inadequate student achievement and preparation in mathematics and science (National Academy of Sciences 1982, National Science Board 1983). One of the most often articulated rationales and visible influences on educational reform policy for more than a decade has been the state of the economy. Indeed, several groups representing the private sector also produced reports on the inadequacy of students' educational achievement in view of what they perceived to be the changing demands of the workplace (Task Force on Education for Economic Growth 1983, Twentieth Century Fund Task Force 1983).

The many reform initiatives subsequently undertaken have embraced seemingly divergent policy directions—some embodying centralizing tendencies, others stressing the decentralization of educational decisionmaking. The enunciation of national education goals, state mandates for more academic course taking, the formulation of national standards, and national and state assessments of student learning are reflections of the former; the decentralization of large urban school districts, site-based management of schools, and teacher professionalization and empowerment reflect the latter, as do efforts to privatize education through vouchers to individual families.

However, before discussing further some currents that have given the present reform movement in science and mathematics education its special characteristics, we need to set this movement in historical perspective. First, reform initiatives have characterized U.S. education for well over a century. And while each of these initiatives has been motivated somewhat differently and has emphasized a variety of specific goals and processes for achieving them, there is continuity in some of the underlying concerns as well as in proposed means for improving prevailing educational practice. For example, conceptual understanding of the underlying structures of mathematics and of major scientific principles, a hallmark of some current reform efforts, was an aim of such new mathematics and science curricula of the 1960s as the texts produced by the School Mathematics Study Group and the Physical Science Study Committee. Present-day reformers would argue that their approach is different, lodged not in intellectual abstractions but rather in student experiences and in practical applications of interest to them. However, this philosophy, too, has a long history dating back to Dewey and his emphasis on project work (Cremin 1964), as we note in the next chapter.

Second, some of the very perceptions of shortcomings within U.S. science and mathematics education—inadequacy of the current curriculum and instruction to prepare students for citizenship and work in the 21st century—are shared by other industrialized countries and have motivated them to undertake reforms aimed at these subjects. One indicator of the worldwide interest in mathematics and science education is the participation of some 50 countries in the Third International Mathematics and Science Study (TIMSS)—an unprecedented international undertaking to compare student achievement in these fields and the educational conditions that may account for observed differences; a second is the set of 15 case studies carried out in parallel with the eight U.S. studies by countries participating in the Organisation for Economic Co-operation and Development (OECD) project and the subsequent conferences on three continents to disseminate their findings and discuss implications.

The point is that educational innovations in science, mathematics, and technology do not originate and evolve in isolation. They are shaped both by macro and micro forces that have currency at the time and by a country's educational history and culture, by the same forces that influence all the political and social institutions of a country, and by forces that operate at a given time within subgroups—be they education reformers, technologists, researchers, teachers, or individual schools. This is amply illustrated by the case studies reported on in this book: Each shows the influence of these forces, though a reading of the individual cases reveals a very different mix of local conditions, national policy directions, and even international influences shaping the initiation and progress of the innovation.

Why Reform?

Contemporary social perceptions and resulting political movements have the power to affect fundamental features of the school—how it is structured, how students are to be engaged, and how teachers carry out their responsibilities.

And, with respect to curriculum, many of these society-wide influences affect not just how mathematics, science, and technology are conceived and taught, but all other school subjects as well.

National Policy

While several of the current centralizing policy directions may be distinctly different from earlier reform approaches, the emphasis on mathematics and science education recalls the efforts of the 1960s. The motivation for improving learning in these fields, however, changed considerably in the 1980s.

Science and Math for All. The 1960s reform efforts gained momentum largely as a result of the launching of Sputnik by the USSR in October 1957, even though some curriculum and teacher training projects were initiated before that event. Sputnik posed a technological and military challenge to the United States, the response to which was the development of a scientific workforce capable of overcoming and surpassing the Russian achievement. Hence the emphasis of 1960s efforts, at least initially, was on developing and increasing the scientific talent pool, that is, providing up-to-date curricula and well-trained teachers in order to ready a greater number of students for majors in the sciences and mathematics and to motivate them to continue into graduate school and scientific careers (Raizen 1991). By contrast, the current wave of reform is motivated by economic and social challenges, and is aimed at *all* students. The goal for the 1990s is science and mathematics for all, as exemplified by the titles of some of the publications developed by the projects we studied—for example, the American Association for the Advancement of Science's (AAAS's) *Science for All Americans* produced by Project 2061 (AAAS 1989) and the Mathematical Sciences Education Board's (1989) *Everybody Counts*, associated with the mathematics standards developed by the National Council of Teachers of Mathematics (NCTM).

Most commentators ascribe the political and financial support for the renaissance in science and mathematics education to several factors:

- the need for the United States to maintain its competitive position in the world economy, which is seen as increasingly dependent on technology-intensive industries and services and, therefore, as requiring a workforce—beyond and quite apart from Ph.D. scientists—adequately educated in science and mathematics;

- the need for the average citizen to know and understand enough to deal in an informed way with individual, family, and community decisions increasingly linked to science and mathematics—for example, concerning health, environmental issues, and technological developments; and

• in the face of these perceived needs, the low standing of U.S. students in a number of international assessments that have compared their mathematics and science achievement to those of students in other industrialized nations (IEA 1988, Lapointe et al. 1989 and 1992), as well as the decline throughout the 1970s of student achievement within the United States (Mullis et al. 1991).

Education and the Global Economy. At the national level, economic concerns appeared to motivate almost every report on the status of education. For example, much of the justification for the sense of crisis that permeated *A Nation at Risk* centered on the perception of America's declining industrial competitiveness. Issues of unemployment, productivity, and the balance between exports and imports seemed to point to a troubled economy. As noted, the implicit linkage to education was fostered and reenforced by reports on education and the economy issued by associations representing business and industry. Such concerns led policymakers and the public to concentrate on the connections between economic vitality and what was happening in school: whether students were being provided with and learning from curricula adequate for new demands, whether teachers were capable of bringing all students to acceptable standards of knowledge and performance, whether schools were organized to facilitate rather than impede the teachers' job, whether parents and communities supported the curricular and educational goals seen as necessary for preparing students for a changing economy. Policies and reform efforts originating in concerns with the country's economic performance emphasize a highly instrumentalist view of education, rather than knowledge seen as a good in itself—liberating for individuals and an indispensable civic virtue in a democratic society.

One diagnosis given for weaknesses in the U.S. economy is the shift from the industrial age to the information age. The thesis is that manufacturing will continue to move to cheap labor markets; developed economies where labor costs are high, as is the case in the United States, will have to turn from a manufacturing to a service economy requiring a highly educated workforce that can develop and provide sophisticated services based on information technology to the domestic market and to other countries (Commission on the Skills of the American Workforce 1990). Hence, the cure for what seemed to be the lagging success of the United States in the global marketplace a decade ago was to educate future workers in mathematics, science, and technical fields so that they could contribute productively to the revolutionized economy.

Economic concerns and the associated perception of inadequate educational achievement clearly underlie three of the six national education goals established by the president and the governors in 1989 (National Governors' Association 1990) which have formed a cornerstone of education policy for two administrations, even though they were drawn from different parties. These goals are:

- Goal 3: Students leave grades 4, 8, and 12 with demonstrated competency over challenging subject matter (including mathematics and science), so they would "be prepared for responsible citizenship, further learning, and productive employment in our . . . modern economy".

- Goal 5: U.S. students are "first in the world in mathematics and science achievement".

- Goal 6: All Americans possess "the knowledge and skills necessary to compete in a global economy and exercise the rights and responsibilities of citizenship".

Coupled in this way with education, the economic motivation has been driving many features of current educational reform. It has steered curriculum in several different directions: the emphasis on the "basic" academic core remains, but this has been redefined to include scientific and computer literacy for all, often expressed in terms of increased high school graduation requirements. In addition, course content is being changed to address the problem of under–preparedness. These changes, too, take several different forms, sometimes striving for integration across traditional subject matter, sometimes for coordination, sometimes for embedding concepts in real-world contexts and issues. Teaching strategies are the subject of reform as well, because traditional mathematics and science education has proved ineffective for large segments of the student population who will be part of the future labor pool. Employers have added demands that prospective workers acquire effective communication skills, appropriate work behavior and ethics, an ability to work effectively in groups, and an interest in the practical applications of school subjects. These demands, too, are having their effect on school reform, originally in the form of integration of academic and vocational education, with a stress on more rigorous science and mathematics in vocational tracks; and more recently in school-to-work programs emphasizing links between school learning and work experience.

Even as these reforms were being initiated and were far from fully implemented, the U.S. economy began to recover during the early 1990s. This recovery—probably quite rightly—has not been seen as an outcome of any of the reforms. This seems sensible enough, given the incommensurate time scales between educational reform and economic cycles. On the other hand, no one has questioned the original analysis linking the state of the economy with the state of education, nor have concerns with the unsatisfactory performance of the schools eased in the least.

Scientists and Educators

While the economic motivation has loomed large at the national and state policy levels, the scientific and educational communities have had their own perceptions of the needs for reform in mathematics and science education: the changing character of the fields of science and mathematics brought about by the advent of powerful computers; the disaffection of much of the student population with science and mathematics courses; the accumulating research on how children learn, specifically in mathematics and science; and the impetus for alternative curricular and instructional strategies provided by educational technology.

Changing Views of Science and Mathematics Learning. Even as the demands increased for better links between the needs of the economy and the education provided to students, changes were taking place in the epistemologies of science and mathematics as well as in the very nature of the what and how of scientific and mathematical work. Sociologists and historians of science and mathematics developed a body of scholarship illuminating these fields as social constructions of the communities engaged in a particular scientific or mathematical field, involving agreed-to forms of methodology, evidence, and discourse. These epistemological views of science and mathematics have influenced and supported changing views of how these subjects are learned and therefore of how they should be taught. The learner is conceived as an active participant in constructing meaning from his or her surroundings (including from the learning opportunities presented in school) rather than as a passive absorber of information. Although not always enunciated specifically, the "constructivist" view of the learner permeates most current educational reform. It is prevalent as well in our cases, though not always carried through in project materials or in the classroom.

At the same time that scholars were redefining the nature of scientific and mathematical knowledge and methods, computers and other information technologies were changing not only the kinds of problems scientists and mathematicians were able to address but also the ways in which they addressed them. More often than not, current scientific research cuts across several subspecialties and involves teams doing different but related tasks and staying in active communication. This research mode is reflected in the Human Genome project, which is devoted to identifying the many millions of human genes, or in research papers in high-energy physics that may list a hundred coauthors. What is remarkable is that demands for an education that better prepares young people for the future workplace, changed conceptions of the learner, and the changing fields of mathematics and science all seemed to point toward the same reform directions, each of which is mirrored in our cases to some extent:

- more learning through real-world problems and applications in context rather than through abstraction,

- making connections across fields of science and mathematics and to other subjects as well,

- the use of computers and other technological aids to work on investigations and problems previously inaccessible to students,

- learning to work in groups, and

- learning to communicate in language and symbolic forms appropriate to the topic.

These themes are further elaborated in the next chapter, which takes up in greater detail changing conceptions of science and mathematics teaching and learning.

The Changing Student Body. These reform directions were further reenforced by the changing nature of the school population. One aspect of this change is easily documented: The makeup of the K-12 student body is becoming increasingly diverse. For example, according to California sources (Catlin 1986), minority students constituted 27 percent of the 1970 California school population. By 1980 the proportion of minority students had risen to 42 percent, and by 1990 it was estimated to be reaching 46 percent. By 2000, minorities in California public schools are expected to surpass nonminority students, comprising 52 percent of the school population. While California may be setting the pace, the rest of the nation is not far behind. The percentage of black and Hispanic students enrolled in grades 1-12 increased from 42 percent to 52 percent during the two decades between 1970 and 1989 in central city public schools, and from 10.6 percent to 20 percent in other metropolitan schools (NCES 1992). Added to these increases of students from minority groups are increases of students whose native language is something other than English. Catlin (1986) estimates that 23 percent of the 5- to 17-year-olds in California public schools speak a language other than English at home. Students considered to have limited English proficiency comprise 11.9 percent of all elementary and secondary school students, with over 40 different languages spoken in the students' homes, ranging from various forms of Spanish and Chinese to Tagalog and Hmong. Nationally, between 1980 and 1990, the number of children ages 5-17 who spoke a language other than English at home increased from 4.5 million to 6.3 million, or from 10 percent to 14 percent of all children (NSF 1996, p. 5).

Past experience of the schools and the underrepresentation of African-Americans and Hispanics in the scientific and technical enterprise indicate the failure of traditional teaching methods in science and mathematics to interest most students from minority populations in these fields or to prepare them to continue, even if they evince any inclination to do so. The increasing cultural diversity of the student body in U.S. schools—shared to a lesser extent by some of this country's European trading partners—has forced a reexamination of pedagogical approaches and a search for more effective ones that will lead students to engage with the content and processes of science and mathematics.

The other aspect of the changing nature of the students now in school is harder to document but equally real to those who meet them face to face every day. It concerns the cultural and societal changes that mark all advanced industrial and post-industrial societies: the changing nature of family and community, the universal access to sophisticated news and entertainment media, the speed of communication, the influence of peers, and rapidly changing fashions. The effects on young people are pervasive throughout the OECD countries: teachers in Germany, England, Japan, and the United States alike comment on the fact that their students would never sit still for the kind of instruction that marked their own school days a generation earlier. Thus, the need to engage students through more relevant curricula and effective teaching strategies is great for all students, and is especially urgent for students from ethnic and language minority groups.

The changing nature of science and mathematics, conceptions of the learner, and makeup and disposition of the student body have contributed to a reformulation of what science and mathematics is worth knowing and how these subjects are to be taught. As a result, there is decreased emphasis on traditional disciplinary knowledge and greater emphasis on linkages across subjects, issues-related topics, and mathematics related to real-world problems—reenforcing the reform directions shaped by the need to prepare a more ethnically and culturally diverse school population for the workplace of tomorrow. Because the changing views of the fields of science and mathematics and changing views of the learner are so significant in the educational reforms of the 1980s and 1990s—not only for the U.S. cases but also for those of the other OECD countries participating in the international case study project—we devote the next chapter to a fuller discussion of how these changing views have shaped the innovations we have studied.

Strategies for Reform

When attention to the state of science and mathematics education reemerged in the early 1980s after having been displaced in the 1970s by other educational priorities, a question for many was why the 1960s reforms had "failed"—why a new round of reforms was needed. Indeed, some of the early investment strategies of the 1980s resembled key strategies of the 1960s reforms, stressing development of improved curricula and opportunities for teacher development. For example, in 1987 the National Science Foundation (NSF) launched the Triad materials development projects aimed at improving elementary science education through collaboration among curriculum developers, schools, and textbook publishers. As another example, for fiscal year 1993, federal agencies requested $436.5 million for development of science and mathematics teachers, mostly aimed at enhancement of teachers already in the schools (FCCSET 1992); from 1993 through 1998, 600,000 teachers were to "receive intensive disciplinary and pedagogical training through Federal agency teacher enhancement programs" (FCCSET 1993, p. b17). Both materials development and teacher enhancement programs continue to be mainstays of federal support for science and mathematics education reform.

Yet there are important differences occasioned by the changed goals and the lessons learned from experience with the earlier reform initiatives. These lessons, discussed in greater detail in chapter 5, concern the need to coordinate activities across potential intervention strategies instead of supporting each in isolation, to build a common vision of what makes for good science and mathematics education, to enlist the various stakeholders in partnerships (as exemplified by the Triad projects) so as to facilitate needed changes, and to create some indicators and accountability measures for improvements. As the emphasis shifted from individual curriculum reform and teacher development toward support of the development of national standards, state curriculum frameworks, and systemic initiatives, reform strategies tended toward centralization, even as the roles of teachers in innovation and reform were increasingly recognized and fostered in some projects.

Standards

Although contrary arguments have been advanced (Bracey 1996), the general perception has been that contemporary U.S. students perform poorly in mathematics and science compared to their counterparts in other countries or even to earlier cohorts of U.S. students. One response has been the development of standards for what students should know and be able to do in a given curriculum field. The apparent consensus in the early 1990s that standards be developed nationally, although eroding somewhat after half a decade (Ravitch 1995), is a radical departure from the strong localism that is a hallmark of U.S. education. Localism is still honored, however, even as the standards movement has gained momentum: Standards are intended to present a vision of curriculum and teaching in a given field and to provide criteria for state and local school personnel so they can decide on the specifics of what should be taught, how it should be taught, and how student learning should be assessed—in short, nationally developed standards but not national standardization. The argument is that standards can be high without curriculum and instruction being the same for all children or sites.

The standards movement has several roots. One is the perception that countries that have strongly centralized education systems, such as Japan and France, are bringing students to higher levels of achievement. In fact, the United States is not the only country formerly relying on decentralized education authorities and policies that is moving toward national consensus about what the schools should be expected to accomplish with their students. England and Wales passed education legislation in the late 1980s which called for a national curriculum and national assessment. Support for such a development, there and here, comes from both ends of the political spectrum. Educational conservatives seem to want government action that moves schools toward the kinds of programs they recall themselves as having experienced and that helped make them successful. They tend to remember their studies as rigorous, challenging, and worthwhile—

a model of what children today should be doing in school. There is also the desire to ensure that money on education is well-spent and that schools are held accountable.

Those who are more egalitarian favor standards, at least in part, for a different reason. They are often dismayed at the record of the educational system in serving (or, rather, failing to serve) large segments of the population, especially the poor and some ethnic minorities. Their attraction to the articulation of education standards, and the impetus this may provide for greater national uniformity with respect to the curriculum, stems from a sense of social injustice. Why should children who go to certain schools receive an education that is far superior to the one students get who go elsewhere? They see standards as a way to raise expectations for all children and make the system fairer.

A potent catalyst for the efforts to develop standards was provided by the National Council of Teachers of Mathematics in their curriculum standards project, the results of which were published in 1989. That project, in fact, is one of the eight cases included in our study of innovations. The NCTM work has consistently been projected as a model for federally supported efforts in other subject matter fields, including science, history, social studies, English, and the arts.

"Standards" is a powerful concept, especially since it encompasses several meanings. It is worth noting that the term has come to be used in the NCTM effort in the sense of a banner, a rallying flag under which professionals in the field of mathematics education could march. NCTM further defines standards as serving three purposes: minimal criteria for quality (originally intended as a check on curricular and textbook claims), an expression of expectations of goals, and means for leading a group toward new goals (NCTM 1989). Thus, NCTM's standards publications encompass statements about education in mathematics that can be held up to exemplify contemporary thinking about best practice. They are intended as much to inspire as to prescribe.

For many of those shaping education policy, however, the idea of standards is attractive because they use a meaning of the term that attaches it to accountability. For them, a standard is a template or set of prescriptions to which objects and activities can be compared to see if they measure up. Their interest in standards is more closely related to making schools, teachers, and students perform at certain levels than to inspiring them. Ambiguity about the term persists and could well lead to controversy when the different meanings begin to clash.

In any event, the development of standards was claiming an increasingly prominent part of the agenda for education policy during the time our case studies were being conducted. Inevitably, the influence of the standards-setting movement is reflected in several of our cases, particularly in AAAS's Project 2061, the California science education reform, and the ChemCom chemistry course, even though all of these were initiated before the advent and high visibility of the national standards-setting efforts. While the issue of the desirability of national standards was seldom addressed directly by these projects, they either influenced or were stimulated to respond to the discussions surrounding the evolution of the national science standards eventually developed by the National Research Council (1996).

Systemic Approaches

Systemic and statewide change is another area of great interest in the reform of science and mathematics education. Systemic change is based on the belief that changing any one component of an education system is insufficient to cause significant reform. Instead, a concerted effort to change the structure, operating procedure, forms of interaction, and distribution of power is needed before any real change in a system can transpire. The experiences with the educational reforms of the 1960s reenforced the notion of systemic reform in science and mathematics education. Given the limited success of interventions addressing a single aspect of precollege education—for example, the curriculum, or the capability of teachers already in the schools, or the preservice education of teachers—policymakers came to the view that these interventions need to be aligned and carried out in concert. Generally included in the policy strategies inherent in systemic reform are the establishment of a coherent reform vision; provision of instructional guidance in the form of curriculum frameworks, instructional materials, and professional development; the restructuring of school governance and organization; and evaluation and accountability mechanisms (Smith and O'Day 1991, O'Day and Smith 1993). Another lesson for current policy learned from the earlier science and mathematics education reform era, which tended to bypass school administrators and state-level personnel responsible for education, is the need to involve the various sectors of the educational system, including not only the institutions and individuals with direct responsibility for education but also the clients—students, parents, and employers.

Based on these premises, NSF has launched a set of systemic reform efforts, involving first states, then large urban systems, and then local rural school districts (NSF 1994a). Federal programs that have as their major focus a specific improvement component, such a teacher enhancement, commonly ask principal investigators and administrators of projects in their purview to set them in a systemic context—that is, to ensure that there are links to other relevant reform efforts affecting participating teachers, schools, and districts (Kaser and Loucks-Horsley 1995). Similarly, the U.S. Department of Education has supported the development of science and mathematics frameworks in 16 states as the base for systemic reform (Humphrey and Shields 1996).

Although California's development of its *Science Framework* preceded these federal initiatives, it illustrates this particular reform direction: The implementation of the *Framework's* vision through the several teacher networks California has set up, as chronicled in our case study, was aided considerably by

NSF support under the State Systemic Initiative (SSI) program. Private organizations, too, have developed comprehensive approaches to reform as demonstrated by AAAS's Project 2061 and NCTM's development of a series of standards documents following the original *Curriculum and Evaluation Standards*. Even when the initiative centers on curriculum development, as is the case for ChemCom and its spin-offs, developers and their sponsors (the American Chemical Society—ACS—in the case of ChemCom) may invest considerable resources in teacher workshops and other dissemination mechanisms, as well as in demonstrating comparability with nationally developed standards and state frameworks and requirements.

Assessment in the Service of Reform

Using standards as the means to elevate the educational enterprise raises two corollary issues: (1) how pupils can be brought, or can bring themselves, to meet the standards; and (2) assessment of whether the standards have been met. The ultimate response to both is assessment of student achievement that is aligned with the standards. Particularly when used for accountability purposes—that is, to determine how well schools and education systems are implementing new standards and rigorous curricula—assessments designed and controlled by sources external to the classroom rather than by classroom teachers generally are preferred by policymakers and funders of reform.

External testing of students to establish what they know has a long history in the United States. In the past, such tests have been used for the distribution of educational goods, i.e., for student placement at both lower and higher levels of education (for example, standardized tests used for assignment to various tracks in elementary and secondary schools; Student Aptitude/Achievement Tests—SATs—used for college admission decisions, and Advanced Placement examinations used to award college credit). The amount of student testing greatly escalated in the 1970s, with an emphasis on minimum competency as documented by "objective" measures and the growth of federally funded compensatory education programs that required standardized testing to establish student eligibility.

The changing views of testing—from a means for making judgments about individual students to a policy instrument—can be traced through a history of the National Assessment of Educational Progress (NAEP). Over the years, NAEP's purpose has changed from providing monitoring information on student achievement at a national level to serving as a means for state-by-state comparisons of student achievement. Indeed, some members of its governing board urge that NAEP be further adapted to allow for district-by-district comparisons. Currently, there is talk of using NAEP as a means for assessing the science and mathematics reform initiatives of the last decade. The results of the student achievement tests administered as part of TIMSS certainly will be interpreted as another indicator of the effectiveness of these reforms.

At the same time, reformers have been concerned that the standardized, machine-scorable testing formats in wide use in the United States are providing only limited information on what students know and are able to do in science and mathematics. While such tests, when well-constructed, can provide good information on students' subject matter knowledge and some reasoning skills, they seldom provide information on such generative science or mathematics thinking as the formulation and testing of hypotheses, the identification of a problem embedded in a complex situation, or the development and evaluation of alternative approaches to solving a problem. Nor can paper-and-pencil tests assess simple or complex hands-on performance skills in science (Shavelson and Baxter 1991), or time-limited tests provide information on students' ability to carry out sustained tasks. And because tests are seen by students and teachers alike as the true embodiment of what is ultimately valued and needs to be learned, they exert considerable influence on the curriculum. Hence, as states have developed science and mathematics frameworks intended to incorporate reforms in science and mathematics education to meet new goals and expectations, the demand for alternative assessments more closely mirroring these goals has increased.

Researchers and testing experts have been developing alternatives for use by teachers to appraise their own teaching and their students' learning and—with much greater difficulty—for external accountability purposes in state- or district-mandated large-scale assessments. Often, these efforts have been supported by individual states, as is the case for the California Learning Assessment System (CLAS) (California Department of Education 1995) and the development of portfolio assessments by Vermont and Kentucky. NSF and the U.S. Department of Education also have supported efforts to develop improved assessment methods. Because alternative assessments are much more costly to administer, time-consuming for students and teachers, and more difficult to score reliably, a backlash is developing against their widespread use for accountability purposes, just as has been the case in England and Wales (Black 1994).

Assessment and accountability demands have had a mixed influence on the eight U.S. innovations we have studied. Specific attention has been paid to these issues by NCTM, which has published a special standards volume devoted to assessment (NCTM 1995); California, which invested in the development of its alternative CLAS science assessment (currently at a halt due to lack of funding); ChemCom, which developed a standardized examination for the course; and Project 2061, which commissioned a "blueprint" paper on assessment approaches. On the other hand, neither the Voyage of the Mimi nor the Kids Network includes assessment components in its curriculum packages, nor was student testing of major concern to the teachers involved in the Urban Mathematics Collaborative (UMC) or Precalculus projects. The relatively limited attention to testing and assessment issues, given their prominence at the national and state levels, is discussed further in chapter 6.

Roles and Actors

In the 1960s, reform initiatives in science, mathematics, and technology education largely were the province of academic scientists and the federal agency established to support their endeavors (i.e., NSF). This is no longer true. Improving the scientific and mathematical literacy of the country's students has become everyone's concern—states, scientific and educational bodies, business and industry, and a dozen federal agencies all have become active players in the current reforms. In addition, powerful new communication and computation tools have made new educational approaches possible, even though their potential for science and mathematics instruction is far from realized in the cases we have studied, let alone more generally in American education.

The Role of States

The establishment of the SSI program signals one of the major shifts from the 1960s, namely the importance assigned to the states in mathematics and science education reform. While the federal government began to grant funds through the U.S. Department of Education in the late 1960s to states for compensatory education programs of various sorts, these were largely flow-through block grants allocated on a formula basis, designed on the model established earlier for vocational education. The improvement of science and mathematics education, however, was largely regarded as a federal responsibility and assigned to NSF, since the need for more scientific manpower was deemed to be a national priority.

In the 1980s, these perceptions changed. For one thing, states assumed a much greater role in the financing of local public schools, due to a number of court cases brought to ensure greater equity in school financing. Based on state constitutional doctrines of equal protection and due process (education being a state responsibility), these cases argued that students residing in poorer districts were treated inequitably compared to students residing in wealthier districts and that the level of per-pupil spending should not depend on a district's property wealth. For 25 years following World War II, the states' contribution to public education, averaged over the United States, was around 39 percent, with a local contribution between 55 and 60 percent (depending on fluctuations in the federal contribution of between 4 and 8 percent). Fifteen years later, by 1987-88, states were contributing close to 50 percent and local districts 44 percent; some states contributed well over 70 percent (NCES 1992). With increased funding responsibility came increased interest on the part of the states as to the performance of the schools.

Concerned with their own economic growth, states were particularly interested in the development of a highly educated workforce; improved science and mathematics education formed the road to this goal. Thus, the current emphasis on improving science and mathematics education is prevalent as much at the state level as at the federal level, though the states continue to look to the federal government for assistance in their reform efforts. The interest has gone beyond state officials directly responsible for educational matters; some of the most active proponents of standards for education have been the state governors, who not only formulated national education goals but have publicized NCTM standards as a model of what was needed in science as well as in other school subjects.

The federal government, too, has been emphasizing the role of states in this area, having learned as one of the lessons of the 1960s that a concerted approach is needed that involves partnerships including all the levels of the educational system as well as the private sector. For instance, in addition to NSF's SSI program, the U.S. Department of Education is supporting the development of science and mathematics curriculum frameworks in 16 states and 10 regional consortia to serve state and local education authorities in their efforts to improve mathematics and science education. The Council of Chief State School Officers receives federal support to develop indicators and reports on science and mathematics education in the individual states, and, as noted, NAEP, supported partly through federal and partly through state funds, reports on state-by-state student achievement in mathematics and plans to do so in science.

The Role of Associations

Three of our eight cases—in fact, the most highly visible and most comprehensive—were either directly sponsored by associations representing scientists or mathematicians and teachers concerned with teaching these subjects; a fourth incorporated the work of these bodies in the reform initiative. The largest scientific society in the United States, the American Association for the Advancement of Science, has provided a home for Project 2061 since its beginnings; the American Chemical Society, which not only is the largest scientific society devoted to a single discipline but also draws a great percentage of its membership from the industrial sector, launched the applications-oriented ChemCom project. NCTM initiated and continues to carry forward its standards project, with closely related mathematics reform activities carried out by the Mathematical Sciences Education Board housed within the prestigious National Academy of Sciences/National Research Council.

Although mathematics may be an easier arena for developing standards than other disciplines, all professional associations of teachers provide similar kinds of opportunities for the involvement of teachers in developing curriculum and curriculum standards. The National Science Teachers Association developed the plan for the Scope, Sequence, and Coordination project, which is an integral part of the California science education reform. A fifth project we studied, the Kids Network, found an active distributor in the National Geographic Society for its materials: Distribution would have been difficult to handle for a traditional publisher because of the project's dependence on telecommunications technology.

Professional teacher associations sponsor conferences, committees, and publications; all of these provide opportunities for teachers to be leaders and to have an impact on the field in which they work. As McLaughlin (1990) notes, professional associations can play an important role in supporting teachers who are trying to change their schools.

Scientific bodies were involved in the 1960s curriculum reforms, usually in developing guidelines for undergraduate science and mathematics curricula, but occasionally also at the precollege level; AAAS's Science: A Process Approach (intended for elementary science instruction) is a prominent example. The strong role played by teacher organizations, however, is a new development characteristic of the current era of reform, even though individual master teachers were part of most 1960s writing and pilot teams developing reform curricula in science and mathematics. What is even more notable—and in contrast to earlier reforms—is that these four comprehensive reform efforts initiated by scientific and teacher organizations all were started without the infusion of federal funds, and three have subsequently received only limited support from this source.

The Roles of Private Foundations, Business, and Industry

Private foundations have invested considerably in the current wave of science and mathematics reform, as indeed they did a generation ago. A foundation initiated the Urban Mathematics Collaborative project; another foundation has been the main funder of Project 2061, with additional funding from yet other foundations; a third foundation supported NCTM in its standards work. Less visible in our cases is the involvement of business and industry; this is understandable since much of this involvement is focused locally. Indeed, the formation of partnerships and alliances with local business is most evident in the UMC project, which was aimed at local—though sizable—urban districts. The strong presence of this sector in current efforts to improve mathematics and science education is evidenced by the many partnerships and alliances that have been formed between individual schools and businesses—something rather rare in the 1960s at the precollegiate level. Partnerships and alliances are also a component of systemic initiatives, whether at the state or district level. The Triangle Coalition for Science and Technology Education, itself funded by a private foundation to encourage local partnerships and alliances with schools, counted nearly 170 members in 1994 representing business, industry, labor, science, engineering, and education and with access to more than 700 national organizations and alliances involved in educational reform (Triangle Coalition for Science and Technology Education 1994).

The current interest of business and industry in science and mathematics education reform is based on some of the same perceived economic and social needs that have motivated other sectors. The Committee for Economic Development, encompassing some 200 presidents or board chairs of corporations and universities, published a report a decade ago in which it argues that:

Human resources determine how the other resources will be developed and managed. Without a skilled, adaptable, and knowledgeable workforce, neither industry nor government can work efficiently or productively.

The schools are the central public institution for the development of human resources. Tomorrow's workforce is in today's classrooms; the skills that these students develop and attitudes toward work that they acquire will help determine the performance of our business and the course of our society in the twenty-first century . . .

A firm and enduring commitment to excellence in education on the part of America's business community is not merely a matter of philanthropy; it is enlightened self-interest. As employers, taxpayers, and responsible community members, business can regard an investment in education as one that will yield a handsome return (quoted in Atkin and Atkin 1989, pp. 30-31).

In commenting on the motivation of science-rich institutions to work with schools, Atkin and Atkin (1989, p. 121) conclude that, though corporate motives may be mixed and include elements of self-interest, leaders of science- and technology-linked corporations are exhibiting "a genuine spirit of community service and altruism" and have recognized the special role of industry in improving science and mathematics education.

The Role of Technology

Technological advances not only have changed the nature of science and mathematics; they have enabled curricular innovation and novel teaching approaches to facilitate current reform goals. The advent of the graphing calculator made possible the approach taken by the North Carolina School of Science and Mathematics (NCSSM) Precalculus course to build mathematical learning on the basis of applications. Moreover, one way that the teachers involved in this project kept in touch was through establishing an electronic network. Advances in telecommunications allowed Technical Education Research Centers (TERC) to create software for the Kids Network through which pupils could communicate with students in other schools and with scientists about measurements they had made on similar natural or human-created phenomena. The Voyage of the Mimi combines videos and computer software with print material to present an enticing storyline to students in order to motivate science and mathematics learning. Our case stories also point out, however, the limitations that surround curricula based on technologies not yet standard in most public schools and generally not available in schools serving mainly minority youngsters and those living in socioeconomically disadvantaged inner-city and rural neighborhoods.

Dissemination and Definitions of Success

Dissemination Strategies

The dissemination and lasting impact of innovative curriculum materials produced in the 1960s, particularly those supported with public funds, was deemed inadequate in the long run, even though several dominated the market at the time. One cause was seen to be inadequate involvement of the private sector during the whole course of the development of a new curriculum. Based on this perception, NSF changed its proposal guidelines for curriculum development in the mid-1980s to require partnerships of curriculum developers with publishers and a set of schools, resulting in the Triad projects for elementary and middle school science. Kids Network is one of the Triad projects; its collaboration with the National Geographic Society as distributor of the materials and manager of the telecommunications network was part of the original proposal funded by NSF and is documented in the study of this case. This strategy did not necessarily eliminate problems between developers and distributors, as the Kids Network case indicates. Chapter 5 in this volume takes up the issue of developer-publisher relationships in greater detail. Since its experience with the Triad projects, NSF has again made its guidelines for curriculum development more flexible.

Innovators and developers appear to have become more sophisticated in bringing their efforts to the attention of influential opinion leaders. The investment of having a public relations firm handle the release of the mathematics standards paid off in terms of their high visibility, as detailed in the NCTM standards case study. The publication of *Benchmarks* by Project 2061 was reported by the media as the advent of national science standards even though, as related in the Project 2061 case study, the benchmarking effort had its origin in helping the school teams build alternative models of the K-12 science curriculum. ChemCom's dissemination efforts included not only publication through a publisher well-known for successfully disseminating innovative curriculum materials but an ongoing life for the project within ACS which allowed continuing publicity for ChemCom activities, translation into several foreign languages for use in other countries, and spin-off projects to develop materials using a similar context-lodged approach for an introductory college course and modules to be used in middle school science.

Criteria for Successful Dissemination and Implementation

Reformers, developers, and disseminators may have quite diverse views as to the criteria that should be used for judging what they have accomplished, and all of these may be different from those of policymakers. As reflected in our eight cases, perhaps at one end of the spectrum is the Precalculus project, which started out as a reform of one course in one school. And even though the project developed into a textbook and a network of teachers in other schools interested in teaching a similar course in their classes, the originating teachers are more interested in continually improving their own teaching than in wide-scale dissemination of their work to others. Yet, it is noteworthy that the school in which these teachers are working was founded by an official act of a state legislature expressly motivated by the need for providing improved mathematics and science education for the state's children, and the approach that the course took was inspired by a long-time national leader in mathematics education who was also instrumental in the development of the NCTM *Standards*. And the spread of the course was facilitated by the workshops and conferences for mathematics teachers sponsored by national private and public sources interested in improving the state of mathematics education in the country's high schools.

The objective of a project like the NCTM standards effort is quite different: the intent is that, over time, every school in America will have changed its teaching of mathematics to approach the vision held out by the standards series. From the outset, NCTM ensured broad participation in the formulation of its original *Standards* volume; for its launching, NCTM hired a public relations firm which managed a very successful publicity campaign capturing widespread media and public attention. To further its reform agenda, NCTM foresees a continuing series of standards publications. In fact, two more volumes have appeared since the original one on curriculum and evaluation; the second in the series addresses teaching, and the third student assessment.

The Urban Mathematics Collaborative project was intended to affect each school district that received funding, although teacher participation was on a volunteer basis, and later sites received considerably less funding than the original sites. There is no clear intent that the project is to serve as a model for adoption by other school districts, although the original funding agency sponsored detailed documentation of the project's progress (Webb and Romberg 1994).

ChemCom can be deemed an outstanding success by the traditional criteria of numbers of textbooks sold and classrooms using the course materials; moreover, originally unplanned extensions into middle school and lower division college are under way, based on the popularity of the high school text. As noted, by demand of other countries wishing to use the curriculum, the text has been translated into several foreign languages. A hallmark of the project has been its well-thought-through dissemination strategies supported through use of royalties for teacher awareness and training. It would be inappropriate to apply similar criteria of success to the Kids Network and the Mimi projects, since neither is intended as a full course and both are dependent on the availability of the hardware they use and on an accommodating school schedule. A success question more difficult to answer about these three innovations regards the issue of fidelity: To what extent are the aims of the original developers being implemented in the schools? What adaptations have been made, and are these supportive of or in conflict with the basic objectives of the course or course materials as envisioned by their creators? And, if the reality of the classroom does not exactly match the vision, is the innovation nevertheless an improvement over what students experienced previously? As to policymakers, they are interested not only in wide dissemination and implementation but also in questions of ultimate impact—that is, whether students have learned more science or mathematics than they did previously. These issues are discussed further in chapters 6 and 7, which take up the adoption and implementation of the innovations we studied.

Micro Forces Shaping Individual Projects

We have noted that all eight of the innovations studied were affected in some way, often in several ways, by the major forces currently influencing education and educational reform. At the same time, each innovation's course also has been shaped by the micro forces operating within the unique set of circumstances surrounding that innovation. Though we can identify several of these micro forces, their individual effects are amplified or diminished as they are combined with each other. And while we may be able to draw some general lessons about their individual or combined influences, the specific effects we have observed are very much tied to the particulars of each innovation.

The Role of Teachers

Changes at the policy level of educational reform have been accompanied by a changing view of the role of teachers in educational reform. Because this change emerged so strongly in our case studies, we devote chapter 4 to it; here we highlight some of the context for the changing views of teachers we observed with respect to their role in innovation.

In the 1960s, a few outstanding teachers were involved in each of the curriculum development projects; many more helped try out the materials and, together with university faculty, staffed teacher institutes. Although by far the greatest federal investment was in these institutes designed to increase teachers' science knowledge and, later, their ability to teach new curricula, teachers were not thought of as the initiators or key actors in reform efforts but rather assigned the role of implementers.[1] In many of the current reform efforts, they are seen as indispensable, not only as implementers but as innovators and critical agents without whose energetic and enthusiastic involvement from the very beginning change is unlikely to take place. Two of our cases exemplify this view: UMC was designed by the Ford Foundation on this principle; even more so, the Precalculus course was completely created by teachers and thereby acquired credibility with other teachers. And while ChemCom involved university chem-

---

[1]This should not obscure the fact that a number of teachers involved in the 1960s reforms have become key leaders in the 1980s/90s reforms, a prime example being James Rutherford, who conceived Project 2061 and continues to be its director. Other examples are provided in the NCTM standards case, where several of the key leaders instrumental in conceiving and then writing the standards documents started their careers as classroom teachers during the 1960s reforms.

istry educators, the course was based on needs expressed by high school teachers and involved them both in the development and revision cycles.

The sets of perceptions concerning the need for systemic approaches (with their emphasis on alignment, standards, and assessment to enforce the standards) and the increased role of states sometimes is set in contradistinction to the view that the teacher has to be preeminent in educational reform, with accompanying autonomy and the power to make decisions. This apparent dichotomy is sometimes expressed as "top-down" versus "bottom-up" reform. Yet several of our cases demonstrate that these apparently opposite views can be reconciled in a given reform effort, most notably in the California reform, but also in the case of a teacher organization—NCTM—breaking new ground by developing standards for its field. Another approach to dealing with this apparent dichotomy operated in the basic design of Project 2061, where scientists were asked to define what it means to be scientifically literate, but teachers were to develop curricular models to demonstrate how the literacy thus defined could actually be achieved in the schools.

The UMC study examines in great detail teachers' activities and development in the context of a reform effort that puts them at the forefront in planning and carrying out changes in their teaching. The conceptualization of the UMC project was influenced by the prevailing view in the 1980s that the lack of teacher participation contributed to the demise of the "new math" reform of the 1960s and 1970s. One underlying premise for the UMC project was that empowered teachers, given opportunities for collegial exchange, become more professionally active, more knowledgeable of mathematics, more aware of applications of mathematics, and more conscious of recommendations for change. The expectation was that change resulting from the UMC project would manifest itself at the district level by an enlivened mathematics teaching force, new structures supporting teachers' active participation in making curriculum and school decisions, and the application of some of the latest trends in mathematics education.

The Role of Leaders

It would not be difficult to see each of our cases through lenses focusing on the key leader—the "great man-great woman" theory of innovation. Although such a leader (or, occasionally, several leaders) can easily be identified for almost all the cases, to equate the course of an innovation with the actions and personality of a single person or small leadership group would be a highly oversimplified view. Nevertheless, leadership is crucial to the initiation and subsequent evolution of an innovation. In several of our cases, a single individual is responsible for the vision (either of science or mathematics, the reform process, or both) that provides the rationale for the innovation and creates the organization. Obviously, once the project has been brought to life, the leadership must see to the health and effectiveness of the organization, even as the functions a project undertakes change through its initiation and maturation.

But leadership is responsible for far more. If outside funders require milestones and delivery of products, leaders must be able to meet such expectations while driving toward the reforms at the heart of the innovation. They must be able to sense the changing environment in order to take advantage of new opportunities (as in the case of ChemCom and the Precalculus project) or change course to adapt to new priorities (as in the case of Project 2061's *Benchmarks*) without losing sight of their overall goals. And effective leadership must at times stand back as others who have bought into the vision and the goals take over key aspects of the project, as demonstrated in the California reform and the Precalculus project. All of these requirements must be negotiated successfully to ensure adequate financial support throughout the course of an innovation, including—if so intended—its continuing life in its originating site and dissemination and implementation elsewhere.

Funding Sources

Obviously, outside funding and its level has major effect on the comprehensiveness of a project, its ability to grow beyond its early plans—even its very existence. Again, our cases are studies in contrast: the Kids Network, though TERC may earlier have envisaged a student communications network for common data collection and analysis, was brought into existence through successful competition in a new NSF grants program. ACS was successful in obtaining—in addition to its own support—funding from NSF for its self-initiated proposal to create modules for an alternative chemistry course. The Voyage of the Mimi received more than $7 million in outside funding for the first and second voyages, originally from the U.S. Department of Education, later from NSF; it has been less successful in its subsequent search for support to continue the series and spread its use. The Precalculus project started without any outside funding as an effort internal to the mathematics department of a state-sponsored school focused on science and mathematics. When outside support was obtained, the teachers involved inevitably became more committed to the work they had undertaken, and the project came to take on a much greater urgency to "deliver." The UMC project was the result of a Ford Foundation initiative; Project 2061 has been supported throughout its existence by several major foundations (and only in a its later phases by the federal government).

In contrast, NCTM was unsuccessful in obtaining external funding for its first *Standards* report, but supported the project generously with internal funds. The result was that not only did NCTM receive funding for subsequent work, but the federal government has supported efforts to develop standards in science, technology, and other fields modeled on the original NCTM effort. NCTM now takes pride in the fact that its landmark *Standards for Curriculum and Evaluation* was the result of its own professional effort. Appropriately, the state has supported the California reform, but the project built on already existing local and regional teacher networks and took on additional impetus with federal funding (through NSF's SSI program) and the National Science Teachers Association's Scope, Sequence, and Coordination project, which has several pilot high school sites in California.

What is startling about the eight U.S. innovations is that all but one were conceived by nongovernmental organizations, the exception being the California reform launched by that state's education agency, and all received considerable funding from nonfederal sources. This "free-market" approach to innovation stands in marked contrast to the initiation and funding of innovations elsewhere: 13 of the 15 cases from the other 12 OECD countries were initiated and funded by the respective national or regional education ministry, the exceptions being the German Practicing Integration in Science Education (PING) project (which subsequently received both state and federal funding) and the Swiss research project on computer modeling. An unanswered question is which mode of initiation and support might lead to more effective innovation—with regard to the nature of the innovation or its adoption—or whether indeed it makes any difference.

Problem Formulation

Clearly, the focus of a project or innovation on what its originators perceive to be the problem(s) will fundamentally structure it. If the problem to be addressed is the need for different curricular opportunities provided to students—e.g., a more issues- or applications-oriented curriculum or a curriculum that takes advantage of available technologies—then the project likely will be more limited in scope than when the perceived need is for a fundamental restructuring of school mathematics or science. Projects may share the view that teachers are at the heart of school reform, but the project will take a very different form when teachers are expected to be the creators of innovative curriculum and teaching strategies than when they are asked to transform someone else's vision into classroom reality.

Problem formulation also involves the project leaders' specific ideas of how a scientific or mathematical field needs to be reconceptualized and, hence, how its translation into a school subject needs to be changed. This translation necessarily includes, whether intentionally or not, conceptions of how the subject is learned by students of different dispositions and at different stages in their education. Moreover, project design and structure explicitly or indirectly reflect assumptions about how school improvement takes place and can be facilitated, what can be accomplished over a given time span, where the levers of change are, and who must be involved inside and outside of the formal school structure. Together with the conceptions of science and mathematics embodied in the eight innovations we studied and discuss further in chapter 3, we consider each project's conception of education reform so fundamental to its whole course that we devote an entire chapter (chapter 5) to this topic.

Project Organization

Even when projects have relatively similar views on such key matters, the project's organization will influence the roles of everyone involved, no matter how integral or peripheral to the project that role is. The degree of centralization, the relationship of the "inner core" to others working on various aspects of a project, the configuration developed for dissemination and implementation, the position of program evaluation and student assessment—all will affect the work and therefore the results of a project. For example, in the cases studied, the hub-and-spokes configuration may be contrasted with a network organization, the deliberate pre-development choice of a publisher to an after-the-fact selection, a multiple-committee structure with a strong central staff core.

As with the origination and basic formulation of a reform project—how close to the classroom, or to the educational governance structure, or to experts residing outside the school—the organization of a project often is classified a priori as top-down or bottom-up. We find this facile dichotomy misleading. A case in point is the California reform, which is based on an initiative by the state's department of education (top-down) but which uses as its major reform vehicle extensive teacher networks with considerable autonomy (bottom-up). Another illustration of an effective mixed strategy is NCTM's development of the mathematics standards, carried out through committee arrangements with an apparently diffuse authority structure yet continuity in widely shared leadership. An example of a deliberately planned mix of organizational arrangements is provided by Project 2061 with its central office, the initial committees of scientists defining their disciplines, and the six participating school districts; yet, the central office clearly wields considerable authority. Thus, within a given organizational structure, the leadership likely will define relationships among the operating components, including the degree of central or localized decisionmaking, how well the different components work together, and how effective the organization is in carrying out its goals.

◆

Our eight case studies set forth in detail how the macro context and the micro currents have helped shape the course of each innovation; we urge interested readers to consult Volume 2 (for the science cases) and Volume 3 (for the mathematics cases) of the *Bold Ventures* series for the fuller stories of each innovation. In this book, we have attempted to look across the innovations. In the next four chapters, we take up in detail some of the themes briefly noted in this chapter: changing conceptions of science, mathematics, teaching, and learning (chapter 3); changing roles of teachers (chapter 4); and changing conceptions of reform (chapter 5). Chapter 6 discusses several elements of reform we found surprisingly underplayed in some of the innovations studied. The last chapter (chapter 7) presents a cross-case analysis of adoption and implementation. We close with some summary reflections (the Coda) on the innovations and about our study of them.

3

The Changing Conceptions of Science, Mathematics, and Instruction

J. Myron Atkin
Jeremy Kilpatrick
Julie A. Bianchini
Jenifer V. Helms
Nicole I. Holthuis

The idea that students should investigate practical matters of importance to the local community is hardly new. Neither is the belief that disciplines should be linked and taught in an integrated fashion. In the 1930s, science textbooks had titles like *Everyday Problems in Science*. Students learned about soil erosion and conservation methods; about infectious disease, vaccination, and inoculation. In mathematics, they learned about consumer discounts, compound interest, and balancing checkbooks. More than three decades earlier, John Dewey pioneered an educational philosophy of focusing school work on neighborhood problems and drawing from the many disciplines as students engaged in such activities. He pointed out that real-world problems are not readily confined within disciplinary boundaries, and neither should schooling be so confined.

Thus, seeing these approaches to science and mathematics emerge as strongly as they do in virtually all of our studies is no surprise to those with long enough memories or a reasonable grasp of curriculum history. For those without the recollections or historical knowledge, however, these shifts in science and mathematics education may seem more remarkable. For the last 40 years, the strongest impact on the science and mathematics curriculum has come from academic scientists: university-based researchers who, in their professional pursuits, are often trying to extend fundamental understanding of their fields. The content they have promoted for study in grades K-12 has been that which they see as fundamental to the development of their discipline, not necessarily that which is most relevant to students or society.

In this chapter, we examine some of the specifics of the shift toward the practical and toward subject integration that is so clear in almost all of our eight studies—and indeed in the case studies of innovations in the other 12 countries that participated in the Organisation for Economic Co-operation and Development project. We look at the forms of this change and highlight what students are studying in the innovations described here. But first we present a framework that may lend some coherence to what otherwise may seem a puzzling set of changes. Educational developments cannot always be understood unambiguously, but we believe there are forces today that are particularly potent in accounting for what we found in the case studies.

Changes in the Fields of Science and Mathematics

School science and mathematics rest on several foundations: curriculum traditions, the knowledge and interests of teachers, the expectations of the public, and the requirements of college admission and of the workplace. They also depend on two elements that we accent here: first, the ever-changing nature of *what* scientists and mathematicians actually spend their time studying and, second, *how* they go about conducting their inquiries. Both of these curriculum foundations for science and mathematics appear to be changing markedly in the late 20th century.

The public expects science to change. There are new facts, new concepts, more powerful explanations for what we see in the world around us. In addition, science is now seen as a field subject to social influences as well. Science has its internal rhythms driven by the logic of its inquiries, to be sure; but it also is influenced by the times and—particularly nowadays—by what the public is willing to support. As an example, some physicists are dedicated to probing the fundamental nature of matter; however, the public, through its elected officials, says that the practical payoff of such research, which is expensive, is not worth the price. There are many problems that seem to call for technical and scientific effort with much greater potential impact on the electorate: finding cures for dangerous diseases, developing less costly methods of protecting a sometimes fragile environment, designing better products and producing them more efficiently to enhance the country's economic position.

The public is less aware that the mathematical sciences are changing, too. Although accomplishments such as the resolution of the four-color problem and the proof of Fermat's Last Theorem occasionally appear in the news, the popular image of the mathematician has remained close to the idea of someone who does a lot of calculations—today with the aid of a computer—rather than someone who creates new knowledge. The computer's impact on the *substance* of the mathematics that is done today is all but invisible to most people, even to those who can see its impact on *how* that mathematics is done. What is often not appreciated is that the growth of technology, new applications, and developments within mathematics itself have greatly extended the scope of the mathematical sciences over the past half-century. Problems in economics and the social sciences, as well as large-scale problems in engineering and the natural sciences, used to be unapproachable through mathematics. Advances in high-speed computing have opened up new lines of research and new uses of mathematics, however. The mathematical sciences now combine recent findings from traditional areas of research such as algebra, geometry, and number theory with methods from applied fields such as operations research, statistics, and computer science. With the greater attention to applications within mathematics itself has come a stronger realization that, like science, mathematics is embedded in a social matrix. More than in earlier decades, the natural and mathematical sciences are understood as human activities and products, and scientific directions and knowledge as dependent on external as well as internal factors.

However much science and mathematics are socially framed, however, they are also bodies of reliable knowledge. And so how to help students understand what science is like today, as well as what scientific knowledge is more or less secure, is the educator's continuing challenge and an underlying theme of all the innovations reported in this volume.

Changes in School Science and Mathematics

The innovations in science and mathematics education in the eight American case studies share a common goal: to define the "essence" of science or mathematics and then figure out the best way to teach it. Each of the projects, however, interprets the essentials in a different way.

There are commonalities among the science-related projects, to be sure. Each makes mention of relationships among the separate science disciplines; each emphasizes the importance of inquiry (though to varying degrees); each, to some extent, accents the point that science is a human endeavor. On close examination, however, the differences among the projects are far from trivial. Those who wish to be informed about these projects—teachers who are developing curriculum, high school science departments that are considering the purchase of instructional materials, textbook publishers trying to decide on priorities—have some fairly clear choices when it comes to the picture of science they want to convey to students.

The mathematics-related projects, too, have important commonalities. All emphasize the links of mathematics with its applications and play down or eliminate the attention to the "structure of the discipline" that was a hallmark of the new math era of the 1950s and 1960s. All of the projects also reflect in some form the continuing attempt by mathematics educators to co-opt the rhetoric of those calling for a curriculum that goes back to basics. The strategy is to redefine what the basics are. In the 1980s, problem solving was defined as one of the basic skills needed by all students. There was also much discussion of computer literacy. Today, the talk is more of a core curriculum and of the need to develop a new vision of mathematical literacy for all students (for details, see the National Council of Teachers of Mathematics—NCTM—standards case study in Volume 3 of the *Bold Ventures* series), although problem solving remains an important theme. By embedding this talk in the rhetoric of "standards," the mathematics education community has enlisted substantial support from politicians and the public.

Literacy is a relatively new concept as applied to science and mathematics. It is also somewhat fuzzy. It seems to have come into educational currency largely for advocacy purposes during the popular movement in the 1970s to move education "back to the basics." Science and mathematics educators were afraid that a sharp focus on the "fundamentals" would result in devaluing their subjects, and so they claimed that science and mathematics were as fundamental as computation, reading, writing, and speaking: one has to be "literate" in science and mathematics, too. The matter of what this literacy is for, however, how it links to the kind of personal and economic survival associated with reading and communicating is still underdeveloped and underexamined.

The rationale for science and mathematics literacy runs a broad gamut that includes equipping individuals for the challenges of the workplace of the 21st century and thereby improving economic productivity; helping individuals with personal decisions such as selecting a healthful diet, coming to grips with one's sexuality, and choosing whether to take up smoking; leading students to an understanding of the elegance of scientific thought; providing a basis for career selection; and gaining some of the tools needed to be a responsible member of the community.

An Applications Orientation

With respect to science, more applications than previously are sought and taught in almost all of the cases we have studied. In most of the innovations, however —ChemCom being an exception—the impression is conveyed to students that science is studied first, then its relevance in addressing issues of practical consequence is determined. This view is not uncommon among scientists, though it is being called into question increasingly as we come to understand better the complex relationships between science and how people make use of technical knowledge. Often we learn to do things better before we understand the principles involved: The technology can precede the science and be a stimulus to it. For decades, fabricators were crafting new metallic alloys to meet industrial needs without knowledge of the underlying scientific explanations. Similarly, sanitation procedures were instituted in hospitals before there was a germ theory of disease.

With respect to mathematics, a similar shift to more practical work has been seen in recent years, and many efforts have been made to bring applications of mathematics into the curriculum. The new math reforms of the 1960s that promised to attract more pupils to the study of mathematics by revealing and employing its abstract structures have given way to the current movement to make school mathematics more useful through application to realistic problems. The curricular compass needle in the mathematics education community has swung around 180 degrees from pure to applied mathematics in just three decades. In the 1960s, there was almost complete agreement that if pupils were taught fundamental notions such as sets, groups, and fields, they would be prepared for the unpredictable uses of mathematics needed later in life. Today, in contrast, there is almost complete agreement that pupils need to see mathematics applied if it is to be relevant either now or later. All the mathematics cases give evidence that the shift has occurred at the level of rhetoric and sometimes at the level of practice.

The story of this shift is a complicated one. Calculators and computers have changed mathematics outside the school and have enabled realistic phenomena to be modeled in the mathematics classroom. Arguments for teaching mathematics to enhance a nation's economic competitiveness have created a climate in which mathematics is increasingly seen as a practical rather than a theoretical subject. A greatly expanded community of mathematics educators nationally and internationally has shifted control away from university mathematicians as arbiters of the mathematics to be taught in school and toward teachers, teacher educators, and other users of mathematics. University-based mathematicians are still consulted by state departments of education and school districts, but their role today is more to ratify the correctness of the mathematics proposed than to provide a rationale, set the themes to be studied, or demonstrate the pedagogy to be used.

Note, however, that most of this change has occurred on the surface. Below the international consensus, at the classroom level where curricula actually live, the influence of the movement may have been felt strikingly—or not at all. It can be argued (Kilpatrick and Stanic 1995, Stanic and Kilpatrick 1992), for example, that none of the so-called reform movements in the United States during this century—neither the unified mathematics proposals of the first decades, the new math movement of the 1950s and '60s, nor the effort to raise standards for school mathematics over the past decade—qualify as true reform of the school mathematics curriculum. In each case, the resulting changes have not been those intended, and the strong rhetoric has masked disunity, contradiction, misinterpretation, and indifference.

The mathematics cases show that the traditional orientation toward mathematics as consisting of routine procedures has been difficult to change. Calls for the integration of algebra and geometry fall on as many deaf ears now as they did at the turn of the century. Attempts to bring calculators and open-ended problems into the curriculum have often resulted in lessons in which these are treated as curriculum objects—here's a lesson on open-ended problems; now we've done that—rather than as tools for learning mathematics. Some teachers have been rethinking their mathematics teaching, attempting to justify curriculum topics according to their utility for applications as the students are introduced to them, and attempting to reduce the attention to procedural topics, as in the Precalculus case study. Many, however, continue to teach a traditional curriculum.

A similar set of conclusions seems reasonable with respect to science. The Stake and Easley (1978) case studies of the late 1970s, more than a decade after the spate of curriculum projects supported by the National Science Foundation (NSF) curriculum projects supported, demonstrated that science teaching, by and large, was textbook oriented rather than investigatory. Recitation was the prevailing mode of instruction in the classes studied, with students expected to respond almost exclusively to teacher-framed questions, preferably using the precise words of the text.

An Integrated Curriculum

The case studies reveal a pronounced move toward coordinating and integrating the sciences—that is, blurring disciplinary boundaries. This development is partly a reflection of the emphasis in the curriculum on practical matters that are difficult to confine within the established disciplines; partly, it is the result of developments within the disciplines themselves that tend to push at the existing separations and lead to the creation of fields like biophysics, medical ethics, exploratory data analysis, and geostatistics.

In mathematics, again at the rhetorical level, there is a vision of "mathematical connections," best expressed in the NCTM *Standards*, as involving links between conceptual and procedural knowledge, recognition of relationships among different topics in mathematics, use of mathematics in other curriculum areas, and use of mathematics in daily life. Again, this is partly connected to the shift to practical applications of mathematics, in which many concepts from science and social science enter the mathematics curriculum that teachers have not had to contend with in the past. And it is partly the result of developments within mathematics itself—and in computer and calculator technology—that increasingly make the partitioning of mathematics into separate curriculum areas such as algebra, geometry, and trigonometry difficult to sustain.

Embedded in these trends is the need to figure out, for curriculum purposes, just how the subjects relate to one another. In science, the discussion centers on the merits of "integration" and "connections." The latter term is often an attempt to preserve existing subject distinctions, but, at the same time, to illustrate relationships among them. The former is frequently associated with entirely new subject sequences and course titles, which are intended to displace the conventional biology, chemistry, and physics. Then there is the use of "themes," which sometimes represents an attempt to bridge these two positions, but probably lies more toward the "connections" orientation. Much of the remainder of this chapter illustrates how these general trends took specific form in the innovations studied. Before providing that kind of detail, however, we elaborate a bit on one more background factor—the general claims for legitimacy of the science and mathematics taught in school.

Sources of School Science and Mathematics

The American Association for the Advancement of Science's Project 2061, with its roots in 1960s-style reform, stands out among the innovations as the only one that justifies its choice of content almost exclusively on conceptions of subject matter advanced by research-oriented professors.[1] In the name of achieving science literacy, the project's guiding question has been: "What should Americans know by the time they are 18?" Relying mostly on the perspectives of the university-based scientific research community, the project defined science as a particular set of concepts, which are gained through certain kinds of practices. According to Project 2061's recommendations for the science curriculum, *Benchmarks for Science Literacy* (AAAS 1993), there are more than 800 separate scientific ideas that a science-literate individual should know and understand by age 18. Furthermore, science is defined in bold and broad terms—unprecedented for purposes of curriculum—to include not only the conventional natural science disciplines of chemistry, earth science, physics, and biology, but also mathematics, technology, and the social sciences. The uses to which science is put, while not irrelevant in Project 2061, is considered more of a motivational factor, a pedagogical device, for engaging the students in learning science. Science concepts, not consequences, take priority.

In sharp contrast, ChemCom, while also sponsored by an association of professional scientists (the American Chemical Society) is less wedded to a conception of the field as articulated by its academic practitioners. For this project, the core of chemistry for those in school is to begin to understand how the subject relates to social and political issues that are important in the local community. ChemCom maintains that this orientation to science is best learned by focusing first on a social issue or problem, and then introducing the science concepts on a need-to-know basis. The ChemCom course, therefore, promotes the view that the practical relevance of science is primary for high school students: They should be prepared to participate in community action and learn how science knowledge relates to personal and social issues. Further, ChemCom presents, through its social-issues orientation, an interdisciplinary approach to teaching science. The developers believe that the various science disciplines (as well as the social sciences, technology, and mathematics) all contribute to an understanding of science-related social issues.

For the California reform project, unlike either Project 2061 or ChemCom, the focus is on a particular organization of the subject matter. The essentials of science are to be found in the conceptual links among the myriad science facts and fields, in the connections that help students understand that science is a coherent and unified field of study. An integrated view of the world of science more accurately portrays the field than a concept-by-concept or discipline-by-discipline presentation.

In the beginning, California promoted cross-disciplinary themes for integrating the sciences. There are six themes in the 1990 *Science Framework*

---

[1] Andrew Ahlgren, associate director of Project 2061, feels this overstates the control of university scientists in specifying subject matter context for the project. See "Project 2061 Comments on the Core Study" in Atkin et al. (1996b, p. 238).

(California Department of Education 1990) text: evolution, scale and structure, energy, patterns of change, stability, and systems and interactions. However, as in Project 2061, which also devotes a chapter to themes in science (and with which the major figures in the California reform were well-acquainted), these them have not proved to be a defining feature of the reform effort. Indeed, themes have emerged as a point of contention and frustration among reformers and teachers alike due to misunderstandings about the use and meaning of the term. (Are "whales" a theme?) Nevertheless, integration of the sciences is perceived by those involved in the California reform as an outgrowth of the theme concept and has become the driving force behind changes at both the elementary and secondary levels. For the elementary schools, in addition, California places a strong emphasis on the processes of science and the power of engaging students in "authentic," hands-on activities.

For Kids Network and Mimi, the processes of science—its methods of inquiry—represent the single most salient feature of the subject. Both projects attempt to bring the doing of science to the classroom through various technologies. This effort, it is believed, will help students understand how science is done in the real world, motivate them to study more science, and help them understand the workings of the natural world in a nontextbook-based fashion. The notion that science is a cooperative process in which all students can engage defines and drives these innovations. For both projects, there seems to be a desire to involve students in activities that mirror the practice of real scientists, either directly, as in the case of telecommunicating in Kids Network, or vicariously, as in the video programs that students watch and discuss in Mimi. In all, the doing of science, wherever the questions under study originate, defines the fundamental feature of science education for these projects.

The NCTM standards project portrays mathematics as a collection of concepts and procedures to be mastered if students are to be productive citizens and literate workers in the next century. The documents reflect a continuing attempt by mathematics educators to convince parents, teachers, and even some mathematicians that school mathematics needs to be more than a sort of enhanced arithmetic computation plus a few other topics such as geometry and proof. To support the argument for reform in school mathematics, the claim is made that technology has transformed mathematics itself, its uses in society, and how it should be taught in school.

The Urban Mathematics Collaborative (UMC) project did not attempt to develop its own unique vision of school mathematics. Instead, it relied on various reform documents, particularly the NCTM *Standards*. Teachers were encouraged to learn about reform proposals but were not pushed in a single direction. Within the collaboratives, teachers' classroom practices reflect an emphasis on applications of mathematics to realistic situations, especially those involving technology; hands-on, cooperative learning in groups; and learning how to communicate mathematically.

The Contemporary Precalculus course developed by the North Carolina School of Science and Mathematics (NCSSM) went beyond the other two mathematics projects in rejecting topics from the canonical curriculum in school mathematics that emerged from the new math era. The course uses applications to introduce concepts to be taught rather than teaching the concepts first and then giving students a chance to apply them. It attempts to portray mathematics as a subject in which students pose problems, collect data, find a model for the data, test their model, make predictions based on it, and analyze its limitations. The course tries to develop students' ability to use the mathematics they have learned—a theme that is common across all three mathematics innovations.

In all eight projects, moreover, along with a greater emphasis on applications and on connecting disciplines traditionally seen as separate, new visions of each subject field have had to be adapted to the realities of classrooms and schools, a topic to which we now turn.

Operationalizing the Views of Science

The nature of science portrayed in the science innovations can be best illustrated by examining how each case approaches the following four issues: integration of the disciplines, the role of "basic" concepts, applications, and science processes. How do the different views of what is important to learn in school science manifest themselves in the various innovations?

Integration: Making Meaningful Connections

Each science innovation—the California reform, Project 2061, ChemCom, Kids Network, and Mimi—addresses integration on some level. Of the five, however, California stands out as the effort that has placed the interrelatedness of the science subjects at the center. At the high school level, where this pattern has had little precedent in the United States, teachers are attempting to restructure their science programs away from the traditional sequencing of subjects toward "integration," where the distinct science disciplines are presented as being so intertwined that they cannot be separated, or "coordination," where each of the four disciplines is treated individually and equitably each year of high school. "Every year, every science, every student" reflects California's effort to reconceptualize when, what, and to whom science should be taught. Science is treated as a core subject, but more to the point of this report, the subdisciplines of earth science, chemistry, biology, and physics, considered separately, are not presented as the main focus; rather, it is their interrelationships, their connectedness, that is to be emphasized. This change has been promoted both at the elementary level and, more innovatively, at the high school level through California's Scope, Sequence, and Coordination project, the most visible effort in the state to reform high school science.

Why this move toward integration? One popular and oft-cited reason is that teaching science as a coherent amalgamation of biology, chemistry, physics, and earth science is more like "real" science; that is, integrated science reflects a more accurate view of actual scientific practice. One teacher puts it this way: "Science, as seen in the real world, doesn't happen in little compartments, does it? You don't see a physics thing happen here and a biology thing happen there. It is integrated in real life, so I think by integrating science we're teaching more like it really is, like it really happens" (Atkin et al. 1996c, p. 58).

Integrating the sciences serves at least two purposes in the new California curriculum: It reflects how many scientists view the field, and it makes science more relevant to the students' lives. Yet the text of the *California Science Framework* (California Department of Education 1990) itself, the guiding document for science education reform in the state, makes no explicit mention of integrated or coordinated science. It does, however, as noted earlier, treat the notion of themes quite extensively—albeit ambiguously—in the content-specific sections of the text.

What does integrated science in California look like? In one Northern California school, all three years of science are integrated into courses titled Science and Technology 1/2, Science and Technology 3/4, and Science and Technology 5/6. These courses are a combination of concepts from earth science, physics, chemistry, and biology. For example, the large units of material for the first year fall into the following categories: Planet Earth, motion, forces and energy, air and water, energy and atoms, chemical reactions, chemical reactions in living organisms, and food and digestion. Within each of these units, the students experience a particular phenomenon drawing on examples from all of the sciences. The section on forces and energy includes lessons on heat transfer and flow, kinetic and potential energy, friction, and equilibrium. Students analyze the evolution of warm-blooded animals to study heat flow, use a pendulum to investigate potential and kinetic energy, and compare the relative grip of the rubber on the bottom of athletic shoes to learn about friction. In a series of lessons on equilibrium, students study models and case studies of how living organisms manage to maintain a relatively steady internal state.

Other schools follow a more "coordinated" curriculum, where physical, life, and earth science are explicitly taught each semester. In one school using this approach, Coordinated Science is a four-semester course designed to meet the graduation requirements of the state of California. Students are instructed with a thematic approach which integrates life, earth, and the physical sciences in a program called "Science in a S.A.C.K. (Spiral Approach to Content Knowledge)." Two of the key features in this program are exposure of students to multiple disciplines of science in a given unit of study and the fact that no one topic is studied to a maximum depth in any one semester. Instead, the topics are spiraled throughout the four semesters—introduced one semester, then expanded upon sequentially in succeeding semesters. In the first semester of this program, students study the overarching topic of water quality. Through the use of discussions, labs, other hands-on activities, and lecture, students learn about the physics associated with water, such as molecular structure, static electricity, and super-freezing. They also study how water affects the earth through the formation of rivers, aquifers, and aquitards. The chemical concepts addressed in this unit are surface tension, chemical nomenclature, water purification, symbols and formulas, and the periodic table. Biogeochemical cycles, biomes, viruses, and respiration in invertebrates and fish are examples of the biological topics.

Both of these examples illustrate California's focus on content reorganization. As these examples also illustrate, the reform did not attempt to change the actual content. While the *Framework* offers a bold vision of science as connected through overarching themes, the implementation of themes proved to be problematic, and they were eventually dropped. Integrated science, "big ideas," and "connections" became the new watchwords.

Project 2061, ChemCom, Mimi, and Kids Network also promote integration, although to a lesser degree. In Project 2061, teachers involved with the reform at the various sites gravitated toward an integrated program as a means to implement the tenets of this reform effort. And, as noted, Project 2061, like the California reform, discusses the use of themes. However, the project developers remained neutral on the issue of this particular organization of the curriculum. Notes one Project 2061 staff member,

> I don't think Project 2061 is arguing for integrated curricula. What I think *Science for All Americans* [Project 2061's guiding document] says is that understanding concepts well depends on understanding other concepts . . . [On integration,] the jury is still out. Most of the literate people I know started in a single discipline, and that sophisticated understanding of one discipline made it easier for them to broaden their knowledge in others (Atkin et al., 1996b, p. 198).

On the other hand, ChemCom, while distinctively a chemistry program, makes it very clear that there are important and consequential connections among science, technology, and society. It also integrates chemical concepts with what might traditionally be considered biology, earth science, or physics. To be sure, the aim of ChemCom is to present science as being connected to the students' own lives, which necessitates an explication of where science and society intersect.

Kids Network and Mimi both promote the view that science always takes place in some kind of context. They engage the students in an exploration of a scientific issue that has repercussions in the world outside a laboratory. They attempt to portray science as an activity that happens in many different locations and that is done by many people. Indeed, the Kids Network case report finds that an integrative and interdisciplinary format provided the incentive to adopt the curriculum in most cases. Similarly, the Mimi case report suggests that teachers were attracted to the program because of its explicit attention to the real world and its integrative and thematic format. To this point, one teacher reports: "Kids did not like what we had before: We were studying light waves, sound waves, doing cells, talking about oceanography . . . but it never made any connections. There wasn't a storyline to what we were doing . . . Mimi really makes the connection to the kids' lives" (Middlebrooks et al. 1996, p. 437). Another teacher comments: "It's a perfect place for me to start and jump off from. I just didn't know what integration was until I used Mimi" (Middlebrooks et al. 1996, p. 437).

Forward to the Basics

All of the science cases embrace the notion that science education should be in the business of enhancing students' science literacy, however varied and imprecise that construct may be. Project 2061, however, provides the strongest case for this orientation. The introductory chapter of *Benchmarks for Scientific Literacy* (AAAS 1993, p. xi) states:

> Project 2061 promotes literacy in science, mathematics, and technology in order to help people live interesting, responsible, and productive lives. In a culture increasingly pervaded by science, mathematics, and technology, science literacy requires understandings and habits of mind that enable citizens to grasp what those enterprises are up to, to make sense of how the natural and designed worlds work, to think critically and independently, to recognize and weigh alternative explanations of events and design trade-offs, and to deal sensibly with problems that involve evidence, numbers, patterns, logical arguments, and uncertainties.

The remainder of the text is devoted to explaining the actual concepts and facts that students should know at the end of grades 2, 5, 8, and 12. The chapters parallel those in *Science for All Americans* (AAAS 1989): "The Nature of Science," "The Nature of Mathematics," "The Nature of Technology," "The Physical Setting," "The Living Environment," "The Human Organism," "Human Society," "The Designed World," "The Mathematical World," "Historical Perspectives," "Common Themes," and "Habits of Mind." Within each of these chapters, specific learning goals are described. For example, in the fifth chapter, "The Living Environment," the authors suggest that, by the end of 5th grade, students should know the following about cells (AAAS 1993, p. 111):

- Some living things consist of a single cell. Like familiar organisms, they need food, water, and air; a way to dispose of waste; and an environment they can live in.

- Microscopes make it possible to see that living things are made mostly of cells. Some organisms are made of a collection of similar cells that benefit from cooperating. Some organisms' cells vary greatly in appearance and perform very different roles in the organism.

Each chapter contains statements such as these that cover the wide range of science topics. They are presented according to "grade bands" in an effort to reflect children's cognitive development and are purposefully presented as knowledge statements rather than behavioral objectives. According to Project 2061, knowing something provides flexibility; it "implies that students can explain ideas in their own words, relate the ideas to the benchmark, and apply the ideas in novel contexts" (AAAS 1993, p. xvii). Each benchmark is not intended to exhaust all possible ideas related to it; rather, the benchmarks are considered the core of science learning.

Personal or social relevance, applications of science, a view of scientific inquiry, or a sophisticated vision of integrated science or interdisciplinary science education have their place in Project 2061, but these understandings are to be put in the service of students learning the core concepts. And while several of the six sites associated with Project 2061 are, in fact, engaged in the development of integrated curricula, this is not the chief concern of the key project developers at the Washington headquarters.

The remaining four science projects also present a particular core of knowledge they believe is essential. The bulk of the *California Science Framework* (California Department of Education 1990), for example, is devoted to explicating the specific content that should be taught at the various grade levels. The text is divided into four major chapters: "The Major Themes of Science," "Physical Sciences," "Earth Sciences," and "Life Sciences." Within each chapter, (except the one on themes), specific science concepts are described, increasing in sophistication up through the grade levels.

Each science topic is addressed in the form of a question. For example, in the "Physical Science" chapter, the topic "Energy: Sources and Transformations" is explained through answers to the questions: "What is energy?" "What are its characteristics?" "What do we do with energy?" "What changes occur as we use it?" For each grade level, the questions are answered in the form of short essays that contain what the authors consider to be crucial information for students at that level. Thus, for grades 3-6, the questions "What is energy?" and "What are its characteristics?" are answered this way:

> Energy passes through ecosystems in food chains mainly in the form of chemical energy supplied to each organism by the nourishment it consumes. All organisms convert some of this energy into heat. Animals also convert some of this energy into heat. Animals also convert some of it into mechanical energy. Green plants convert light energy into chemical energy by means of the photochemical process called photosynthesis (California Department of Education 1990, p. 61).

This statement is then followed by a reference within the text for more information on these concepts, and the selected themes that the concept embodies (in this case, Systems and Interactions and Energy). It should be noted that while the *Framework* lists the content of science in separate sections, the authors defend this format by pointing to the narrative form written from the perspective of one or more of the themes. They believe that this format portrays the content of science in a way that "avoid[s] an emphasis on isolated facts and definitions that have long dominated science instruction" (California Department of Education 1990, p. 3).

ChemCom's most outstanding feature is its introduction of concepts of chemistry when they are needed to understand a particular social issue. Nevertheless, the curriculum does embrace the idea that there is a core of chemistry knowledge. For example, the teacher's guide (ACS 1988a) illustrates how and when the program introduces the major topics of chemistry, such as physical and chemical properties, solutions and solubility, nomenclature, chemical bonding, the mole concept, and stoichiometry. Interestingly, one of the present debates over ChemCom centers on the perceived rigor of the curriculum and the extent to which it adequately covers those concepts in chemistry that are representative of the discipline. Electron orbitals, for example, are never discussed, and the mathematics required to solve most of the problems is rudimentary.

Applications of Science

Many programs emphasize the need to make science more personally relevant to today's adolescents and also the need to deal with its applications; yet in most cases the focus is on science first, with issues of practical consequence taking second place.

An exception is ChemCom—a curriculum that attempts to integrate scientific, technological, and societal issues with a focus on decision-making and problem solving. The textbook (ACS 1988b) is divided into the following eight units: "Supplying Our Water Needs," "Conserving Chemical Resources," "Petroleum: To Build? To Burn?," "Understanding Food," "Nuclear Chemistry in Our World," "Chemistry, Air, and Climate," "Health: Your Risks and Choices," and "The Chemical Industry: Promise and Challenge." As is evident from the unit titles, ChemCom highlights the relationships among the sciences and their relevance to the student, the student's family, and the student's community. Each unit stresses the importance of making sound decisions based on all available evidence—some of which is scientific, some of which considers the social consequences of science.

The opening pages of the ChemCom teacher's guide (ACS 1988a, pp. xxxi–xxxiii), list 32 "concepts and assumptions related to ChemCom issues." The majority of the statements contain a strong message regarding the consequences of applying scientific knowledge in the real world. Examples include:

- Citizen participation in public policy decision-making can often be enhanced by understanding the underlying scientific and technological concepts.

- The ability to detect various pollutants often exceeds our capabilities to assess long term health effect and/or to control emissions.

- Citizen involvement in public policy is both a right/privilege and responsibility. Not to decide (or to decide not to decide or act) is also a decision.

- Our present state of knowledge about any given societal/technological issue is likely to contain some imprecision, inaccuracy, and uncertainty. Society must act upon the best available information with the understanding that additional information may call for subsequent re-evaluation of a given issue and/or previous solution.

Every chapter in the ChemCom textbook contains one or more "You Decide," an "inquiry activity where students are presented aspects of societal/ technological problems, asked to collect and/or analyze data for underlying patterns, and challenged to develop and support or refute hypotheses/solutions based on scientific evidence and clearly-stated opinions" (ACS 1988a, p. xxvii). For example, in the opening unit on water, students are introduced to chemistry through a fictitious narrative about a fish kill in a local river that brings the town of Riverwood to a halt. Through laboratory activities, discussion, and problem solving, culminating in a mock town meeting, students attempt to determine what caused the death of the fish and the best possible solution.

As part of this first unit, students study the process of water purification. For one "You Decide," students are asked to determine whether water should be chlorinated. One of the risks of chlorination is the possible formation of substances called trihalomethanes (THMs). One common THM is chloroform, a carcinogen. Given the known benefits and risks of chlorination, students are divided into three groups, each group representing an alternative. They then discuss the following questions:

- Consider the alternative assigned to your group. Is the choice preferable to standard chlorination procedures? Explain your reasoning.

- Can you suggest other alternatives beyond the three given above?

In addition to "You Decide," students may also engage in a "ChemQuandry" or "Your Turn." A "ChemQuandry" is meant "to motivate and challenge students to think about chemical applications and societal issues, which are often open-ended and may generate additional questions beyond a specific 'right' answer." The purpose of "Your Turn" is "to give students practice and reinforcement on basic chemical concepts, skills, and calculations in the context of applied, 'real world' chemistry problems" (ACS 1988a, p. xxvii).

The *California Science Framework* does not directly address the applications of science, except in its discussion of the relationship between science and technology. For example, in the executive summary in a section called "Achieving the Desired Science Curriculum," the *Framework* states:

Science is directed towards a progressively greater understanding of the natural world. Technology is related to science as a human endeavor, but the direction is toward using accumulated knowledge from science and other fields in order to control and alter the way things work . . . Teaching science in the context of STS [science, technology, and society] helps reveal the situations in which science has meaning (California Department of Education 1990, p.4).

It is not clear from this statement, or from the rest of the document, whether the *Framework* advocates the STS approach to teaching science. On close examination, it seems that this reference to STS was made to clarify the distinction the authors draw between science and technology, rather than to promote the teaching of STS. Later, in an explanation of the processes of science, "applying" is described as "a process that puts extensive scientific knowledge to use. Sometimes that knowledge is used in a practical sense as in the building of a bridge. Sometimes it is used to tie together very complex data into a comprehensive framework or theory. And sometimes, the goal is to elaborate and extend a theory" (California Department of Education 1990, p. 152).

The suggestion that application is a scientific thinking process ignores the ChemCom-type definition of application—that is, the occasion to discuss and analyze the personal or social impact of scientific information. California's idea of application is geared more toward the transfer of knowledge within the domain of science, rather than from science to other, social worlds. It should be noted, however, that teachers within the state view the students' ability to apply their scientific knowledge to the everyday world as an important feature of the new reforms; and many programs, particularly those aimed for lower ability students, advocate an applications orientation to the integrated curriculum.

Kids Network and Mimi also exemplify an applications approach to teaching science, primarily through the choice of science topics each innovation presents. One of the principal findings of the Kids Network study was that the "relevance inherent in the Kids Network programs was cited most frequently as an incentive" for adoption (Karlan et al. 1996, p. 312). Examples of Kids Network units include: "Acid Rain," "What's in Our Water?," "Too Much Trash?," and "What Are We Eating?" As with ChemCom, each of these units incorporates science knowledge with how that knowledge can be used. In addition, the Kids Network's telecommunications component represents one way in which scientific knowledge is shared and used by different people.

The Mimi case report refers often to the program as providing a "hook" into science and other subjects, which is primarily attributed to the delivery formats: television, print, and computer software. However, the hook also comes in the form of a story that involves science, social issues, and personal values. Mimi presents science in a context.

As the original vision of Project 2061 was greatly influenced by academic scientists, it treats the application of scientific knowledge as secondary to an understanding of the basic ideas in science and the connections among them. Project 2061 does not directly address science applications. This is not to imply that the project ignores how scientific knowledge is put to practical use. For instance, scattered throughout Benchmarks are statements that refer directly to various practical advances science has made possible. For example, in a section on heredity, the document suggests that students in grades 6-8 should learn that "[n]ew varieties of cultivated plants and domestic animals have resulted from selective breeding for particular traits" (AAAS 1993, p. 108).

The Processes of Science

For decades, educators have debated the role of science processes in science teaching. Is process more (or less) important than content? While one of the 1960s NSF-supported curriculum projects for elementary schools (*Science—A Process Approach*) took a polar position on the issue, and others were predisposed strongly in that direction (the *Elementary Science Study*, for example), most of the projects strove for balance. The issue is still alive in the 1990s. Of the science innovations considered here, the two elementary-level classroom projects and the elementary component of the California project best reflect the view that scientific processes (today usually called inquiry skills) are central to the teaching and learning of science. For example, according to the Kids Network case report (Karlan et al. 1996, p. 253) developers of the project were guided by the following principles in conceptualizing the Kids Network units:

> That students should deal with real and engaging scientific problems, problems that have an important social context. That kids can and should be scientists (students are working as scientists on real science problems). That telecommunications is an important vehicle for showing children that science is a cooperative venture in which they can participate.

This view of science was not necessarily shared by Kids Network teachers, however. Some teachers viewed the program as cookbookish and unauthentic (Karlan et al. 1996, p. 319). Moreover, the focus on processes in Kids Network raised some questions for the case researchers with respect to students' failure to learn accurate science concepts. Whether the activities represented real science or engaged students in authentic experiences, the Kids Network curriculum is centered on hands-on activities. It is the nature of these activities, and their match to actual science experiences, that remains questionable. In the end, the doing of science, wherever the questions under study originate and however authentic the experiences, defines the fundamental feature of science education for this project.

In Mimi, students witness a diversity of people, including a child their own age, engaging collaboratively in activities that they might imagine themselves doing; the developers wanted students to see a more contextual and realistic picture of the scientific enterprise. The case report recounts the thoughts of Mimi's originator on this issue:

> We were comfortable if boundaries around science and scientists got blurry . . . We really wanted kids to imagine a kind of scientific activity that included scientific observation, messing around in the data, looking for patterns . . . We intended to suggest that you could be curious about anything . . . [That] what distinguished scientific curiosity was suspension of belief, questioning of data, challenging of authoritative statements, continuing to keep an open mind about things (Middlebrooks et al. 1996, p. 410).

Teaching Mimi was a challenge for some of the teachers in the study, as it represented a dramatic shift in traditional teaching practices. It required them to "teach adventurously." Further, Mimi emphasizes technology, both as essential tools of scientists and as valuable pedagogical devices for teaching science, which presented some additional challenges. One teacher in the study saw these challenges as an opportunity for teaching a view of the nature of science:

> As far as the kids are concerned, I think they understand what I try to get them to understand: that science is a process, it's not a body of knowledge...It's the process that's more important . . . In order to do this they have to work in groups, teams, cooperative groups . . . There has to be cooperation and there has to be self-discipline among the individuals so that there isn't bickering and fighting and to ultimately work towards a goal. I think that's the main thing I try to get across, and I think they have a sense of it (Middlebrooks et al. 1996, p. 421).

From the teachers' perspective, the case report suggests that many of them felt that Mimi's attention to processes fit very well with their own views of science. Another teacher comment illustrates this:

> There are specific steps that you have to follow in science. Mimi is a nice, gentle way of showing this to kids without saying to them . . . "you must do that." I think at this age that's scary for them. So, I think this is presented in a very positive light, and yes, as the series goes on, there are certain things you have to do. It's like when they go aground. All of a sudden the machines don't work, and something is wrong, and what's wrong? Well, we don't know, but we have to find out, and there are specific ways that they go about finding out (Middlebrooks et al. 1996, p. 422).

In the case of California, the elementary science reform embraced the notion that science processes, taught through direct hands-on experiences, are the key to successful science teaching. The *Framework* is explicit about the role of process skills in teaching science and devotes an entire chapter to them. In the relevant chapter, observing, communicating, comparing, ordering, categorizing, relating, inferring, and applying are proffered as the essential skills for doing science. This chapter also presents a suggested sequence of introducing these skills. According to the California scheme, the first five skills—observing, communicating, comparing, ordering, and categorizing—should be taught by the third grade. The remaining three—relating, inferring, and applying—should be taught in grades 3-6, 6-9 and 9-12, respectively. The rationale for this sequence is based on an unnamed theory of cognitive development.

ChemCom also emphasizes processes in science. However, the processes promoted by this curriculum are the processes of deliberation: ChemCom redefines the essential *process skills* as those needed for engaging in decision-making and real-world problem solving. Knowing the facts is only necessary and appropriate when they help one to understand better the problem at hand; they are not an end in themselves. The traditional processes of scientific investigation, such as those outlined in the *Framework*, are not dismissed in ChemCom; rather, they are deemed secondary in importance given that curriculum's particular vision of what is essential in science education.

Finally, how does Project 2061 approach the role of science processes? Surprisingly, the project does not take a very strong stand on this issue. The program is unwavering on the importance of understanding science concepts, and is also very clear on the need for students to understand how scientists work. However, it does not necessarily believe that students must have direct experience to gain this understanding. Project 2061 key documents offer the history of science, or the attainment of certain scientific "habits of mind," which presumably might be acquired in a variety of ways, as a means to deepen an understanding of science and how scientists work.

ChemCom: A Special Case

Examination of each of the science innovations suggests how each project defines the nature of science and also implies a view of what knowledge is of most worth. ChemCom seems to offer the most fundamental reconceptualization of the nature of science, in that the science is embedded firmly in a social and political context. Project 2061, the California reform initiatives, Mimi, and Kids Network focus primarily on the facts or processes of science; while ChemCom emphasizes the products of science and the effect science has on people's lives and communities. ChemCom neither shuns content nor ignores science processes; rather, it redefines the essential skills as those needed for engaging in decision-making and real-world problem solving. Knowing the concepts of science, ChemCom implies, is only necessary and appropriate when they help to better understand the problem at hand: They are not an end in themselves.

The ChemCom case in many ways epitomizes several key points in the current debates about the purposes of teaching science. Should chemistry be taught at the high school level to produce more chemists? Or should it be taught to "make future voters (taxpayers, consumers) . . . sensitive to the working of the 'scientific mind'" (Rowe et al. 1996, p. 565)? This debate is accented by the fact that ChemCom has been adopted in many schools to serve the lower ability, nonscience, non-college-bound students; that is, it has been designated the lower track science course. The ChemCom developers, on the other hand, most emphatically intended the course for college-bound students. Many teachers argued that ChemCom is in many ways more rigorous than traditional chemistry, but its lack of mathematical orientation and its exclusion of certain core chemical concepts has resulted in the course being viewed as an alternative route, primarily for those unlikely to pursue further studies in science.

Some of the teachers who have a strong chemistry background and experience in teaching traditional chemistry believed that ChemCom does not adequately prepare students for further study in chemistry. Another group of teachers, more flexible in their conception of the discipline, tended to believe that Chem-Com is an important course for majors and nonmajors alike, a course that reflects preparation for living and further learning. The tension between the groups who believe that science should be taught as it really happens versus those who aim to make science relevant continues. The "purists" are seen as being elitist, and the issues-oriented teachers have been accused of "dumbing down" the chemistry curriculum.

Operationalizing the Views of Mathematics

The nature of mathematics portrayed in the three mathematics innovations can be illustrated by examining how each project approached the following three topics: applications, technology, and mathematical processes. How were different views of the nature of mathematics manifested in these projects?

Applications

The NCTM standards project took a middle-of-the-road position with regard to applications. Applications of mathematics were endorsed, primarily through standards relating to problem solving and to "mathematical connections" with other curriculum areas and with students' daily lives, but there were no specific applications or mathematical modeling standards. Moreover, although real-world problems and applications were stressed throughout the *Standards* document (NCTM 1989), there were also standards dealing with mathematical structure and emphasizing deductive reasoning. Relative to the new math projects of the 1960s, the NCTM *Standards* has a much stronger orientation toward applications of mathematics, but relative to the practices seen in the other two mathematics case studies, the NCTM *Standards* document emphasizes pure mathematics quite strongly.

In the schools visited for the NCTM standards case study, applications did not appear to be a topic the teachers saw as especially salient or in need of emphasis. Their emphasis appeared to be more on understanding mathematics than on using or applying it. For example, the mathematics consultant for one elementary school said, "The criteria [of successful instruction] are not algorithmic readiness and multiple choice testing: the criterion is *understanding mathematics*. The experience of problem solving is the means to that end" (McLeod et al. 1996, p. 91). In a high school, algebra and geometry were maintained as separate courses. The geometry course was changed by emphasizing proofs less and geometric discoveries more, but greater attention to applications was not mentioned as a possible change (McLeod 1996, pp. 106-07).

In the Urban Mathematics Collaborative project, even though the vision of school mathematics was nominally that of the NCTM *Standards*, there was more evidence that the practice was moving toward a greater stress on applications. The partnerships with business and industry allowed many mathematics teachers in the collaboratives to see firsthand how mathematics is used in the workplace. For example, one participant in the Milwaukee collaborative's internship program, after spending a year learning about applications of mathematics to business, concluded, "I believe more than ever that teachers need to . . . provide more 'hands on' activities [in mathematics] as well as connect those activities to the real world" (Webb et al. 1996, pp. 280-81).

Another participant in the same program wrote a unit on blueprinting that gave students opportunities to measure, visualize two- and three-dimensional figures, make scale drawings, and engage in proportional reasoning. She had seen, through her experience in industry, how mathematics was applied. She was impressed with the extent to which that application was done collaboratively by teams of workers, and she incorporated an emphasis on communication and on group work into her unit. The Milwaukee collaborative was one of the most successful in helping teachers learn about applications they could use in their instruction, but other collaboratives, too, built partnership arrangements that allowed teachers to see mathematics being applied.

The strongest orientation toward applications was in the Contemporary Precalculus project. The project had begun with the question, "Why are we teaching this stuff?" And the criterion the teachers who developed the course began to apply to any concept being considered for inclusion in the course was "If we can't introduce the concept with an application, then we won't teach it" (Kilpatrick et al. 1996, p. 154). That criterion, together with the emphasis on data analysis and modeling, shifted the course from pure to applied mathematics. The very title of the textbook produced for the project, *Contemporary Precalculus Through Applications* (Barrett et al. 1992), signaled the change.

In all of the schools visited for the Contemporary Precalculus case study, an orientation toward applications could be seen in those courses influenced by the project. For example, one instructor stressed how important the textbook had been in getting students used to an applications approach. The students were not used to applying mathematical ideas and then writing out explanations of what they had done. Teachers in another school pointed out that they gave no attention to proof in their precalculus courses. In these same schools, however, there were many teachers who remained convinced that traditional exercises in proving theorems and verifying identities were essential for developing logical reasoning ability.

Finding an appropriate balance between attention to the abstract structures of mathematics and consideration of the uses to which mathematics is put has been a perennial problem in school mathematics. Recent reforms attempting to tip the curriculum more strongly toward applications have encountered resistance from teachers comfortable with the current curriculum and from university mathematics professors concerned about a loss of attention to mathematics as an academic discipline in its own right.

Technology

All of the mathematics cases placed a heavy emphasis on the role of technology in teaching school mathematics. In every case, students were to be given ready access to computers and calculators as tools in learning and doing mathematics. The NCTM *Standards* document set the tone in its introduction by pointing out that the availability of inexpensive calculators, computers, and other technology had contributed to the shift in industrialized countries to an information society and by arguing that school mathematics needed to reflect that shift. Claiming that technology was changing mathematics and its uses, the authors of the *Standards* (NCTM 1989, p. 8) asserted their belief that

- appropriate calculators should be available to all students at all times;

- a computer should be available in every classroom for demonstration purposes;

- every student should have access to a computer for individual and group work;

- students should learn to use the computer as a tool for processing information and performing calculations to investigate and solve problems.

The authors of the NCTM *Standards* document debated extensively the statement to be made about technology, and especially about the role of the graphing calculator. The preliminary draft circulated to the profession recommended only that graphing calculators be used in grades 11 and 12, but that restriction was later dropped in reaction to feedback from the field. Another point of contention was whether to encourage the use of calculators in elementary school. Although NCTM was on record as supporting such use, responses from parents and some mathematicians to the preliminary draft were negative toward such a recommendation. Nonetheless, arguing that NCTM should lead rather than follow, the *Standards* authors continued and extended NCTM policy on technology use in the final publication.

Reports from classrooms observed in the NCTM standards case study contained relatively little evidence, however, that the schools were using technology to the extent recommended. Calculators and computers were present in the schools, but they did not seem to be employed extensively by students as tools for learning and thinking.

Teachers in the UMC case study, in contrast, reported great changes in their teaching due to the availability of computer software and of scientific and graphing calculators. The technology seemed to be a force that propelled curriculum change by enabling students to have realistic experiences of how mathematics is used in business and industry. Classroom observers in the case study saw calculators present and being used in mathematics classes taught by teachers in the collaboratives. They also saw teachers who were enthusiastic about the changes that technology had made in both the mathematical content they taught and how they taught it. It is difficult to know how widespread this enthusiasm was, but by offering opportunities for teachers to learn about and use new technology, the collaboratives clearly helped advance the recommendations contained in the NCTM *Standards*.

The Contemporary Precalculus project began with several grants to provide the North Carolina School of Science and Mathematics with up-to-date computer technology. When the mathematics teachers saw how students could use technology to graph functions and do various types of data analysis, they were encouraged to move forward to develop a new course. Other teachers were attracted to their work, in part because these teachers could learn from the NCSSM teachers how to use various software programs developed at the school. Although most of the materials were developed before graphing calculators were widely available, the NCSSM teachers were able to incorporate calculator activities into their textbook before it was published and also to produce a companion manual for two of the most common graphing calculators. Subsequent advances in calculator technology were quickly incorporated into the Contemporary Precalculus course, or its equivalent, at all of the sites observed in the case study.

In both the UMC project and the Contemporary Precalculus project, technology served as a spur to change in curriculum and instruction. As the technology changed to allow more possibilities for investigative work, teachers responded by changing their courses further. In all three mathematics case studies, the impact of computer and calculator technology seemed to have been greater at the high school level—particularly the last two years of high school—than in the lower grades. The use of the technology was connected with an orientation toward mathematics as a subject open to investigation and exploration.

Mathematical Processes

School mathematics classes are commonly seen as boring in large part because so much emphasis is placed on the repetition of arbitrary or meaningless procedures. Each of the three mathematics innovations attempt to redefine school mathematics so that it includes a greater emphasis on meaning and understanding and a lesser emphasis on algorithmic procedures. Mathematical processes associated with reasoning, problem finding, problem solving, and modeling are emphasized over routine procedural processes. The first four NCTM standards at each of the levels (grades K-4, 5-8, and 9-12) deal with problem solving, communication, reasoning, and connections. These standards are unofficially termed "process" standards by many in the profession because they cut across areas of content, even though the *Standards* authors did not propose to structure the mathematics curriculum according to a content-by-process matrix.

Observations and interviews in schools attempting to implement the NCTM *Standards* revealed that the document had not managed to shift teachers' and students' beliefs away from an orientation toward mathematics as routine calculation and procedure. Teachers whose classes were observed did appear to be attempting to engage students in mathematical investigations, but there was still considerable emphasis on ritual and routine at the expense of understanding. Secondary school classes spent large amounts of class time on low-level skills. In elementary school classes, some teachers continued to characterize mathematics as essentially computation.

According to the case study report, teachers in the UMC project were, by and large, not seen as having changed their beliefs about mathematics. A "meaningful number" of teachers who were active in their collaborative, however, reconceived their view of mathematics and attributed that reconceputalization to their involvement with the project. One teacher spoke of algorithms and rules as her "pet peeve" and said that even though students want to have rules, they forget them quickly because they do not understand them. She said, "That is what I've gotten away from—these rules in the book—and gotten into the concept underneath" (Webb et al. 1996, pp. 285-86). Processes of inquiry, investigation, and discovery were "prevalent practices teachers associated with their involvement in the collaboratives" and were observed in project classrooms.

The Contemporary Precalculus project began before the NCTM *Standards* appeared, and the rhetoric used by the North Carolina teachers who initiated the project was different from that of the *Standards* document. Nonetheless, much of the same emphasis on thinking and reasoning processes over routine procedures was apparent in the classrooms observed in the case study. Teachers who attempted to incorporate mathematical modeling processes into their instruction, regardless of whether they were using the Contemporary Precalculus materials, almost necessarily shifted emphasis away from routine procedures. A department chairman at one of the sites said that the school's experience with the Contemporary Precalculus materials had helped convince him that the department needed to change the curriculum "from a list of formulas, a list of procedures, a list of skills that need to be accomplished, to a way of thinking" (Kilpatrick et al. 1996, p. 189). At several sites, teachers of the Contemporary Precalculus course had come to see the curriculum not as a list of topics to be covered but as a set of experiences designed to help students see utility in the mathematics they were learning and to be both prepared and disposed to use it. Mechanical skills were taught as needed to solve certain problems; the skills themselves were not the focus of instruction.

In all three mathematics case studies, there was evidence that conventional views of school mathematics held by colleagues, parents, administrators, and students acted as a brake on efforts to redefine the subject as rich in authentic applications, ripe for the use of technology, and rewarding when approached through inquiry and investigation. Teachers who had made such a redefinition themselves were often successful in convincing others, especially when supported by like-minded colleagues, but there were many instances where efforts to change the vision of mathematics were overwhelmed by the status quo.

Changing Conceptions of Teaching and Learning

The changes associated with the disciplines of the natural and mathematical sciences over the past decades resonate with changes in current conceptions of science teaching and learning. Many of these changes have been advocated for decades, at least since the time of John Dewey, if not far earlier; but they have not always been accepted by science and mathematics educators as widely and wholeheartedly as they are today. Teachers have come to believe that students learn in certain ways and that learning is more effective when the students are engaged in work they find relevant. New views of teaching and learning are reflected in three patterns of practice: socially relevant pedagogy, scientifically authentic pedagogy, and learner-centered pedagogy.

A Socially Relevant Pedagogy

In part, the movement to a more integrated curriculum with a focus on "big ideas" is a pedagogical response to what many teachers see as the undesirably remote content of traditional discipline-based courses. As illustrated in several of the cases, both in mathematics and science, many teachers want to serve their students better and have moved toward community-oriented and applied project work, which, in turn, often requires not only a weakening of disciplinary boundaries within the natural and mathematical sciences, but also more explicit attempts to relate science and mathematics to social issues. This development highlights the issue discussed below concerning who "owns" science. It is less clear today than it was a generation ago that science belongs to the university-based researchers. Politicians are taking more control of the scientific subjects and the directions in which they are moving through support awarded or denied for research initiatives. Teachers, too, are redefining the subjects in light of their participation in reform projects and their own views of what is important for their students, as discussed in the next chapter.

For a pedagogy to be considered socially relevant, teachers must engage students in examination of science and mathematics as embedded in and relevant to personal, national, and global issues. As illustrated above, teachers of Chem-Com, for example, are expected to operationalize the curriculum's conceptions of science as enmeshed in society. Teaching is viewed as facilitating activities that promote student collaboration, discussion, and decision-making. In addition to traditional laboratory experiments, teachers of ChemCom ask their students to complete activities such as "You Decide" and "ChemQuandry" which explore issues or applications bearing on the chemistry concepts in the unit. In the Urban Mathematics Collaborative project, the collaboratives build partnerships with local businesses, industries, research institutes, and higher education institutions that show teachers and students alike a variety of authentic applications of mathematics being used by professionals.

A Scientifically Authentic Pedagogy

A second form of instruction advocated by several innovations is scientifically authentic pedagogy—pedagogy that attempts to enculturate students into the concepts, skills, and habits of mind used by scientists and mathematicians. Kids Network provides one example of this approach. To introduce students to the scientific enterprise and engage them in real scientific problems embedded in a social context, teachers of Kids Network guide their students through one or more six-week on-line units. Each unit involves a series of cooperative science experiments in which students use the telecommunications network to send results of their local experiments to a central computer which pools their data and then sends back the combined national results. Participating classes analyze trends and patterns in the national data, examining how their findings contribute to the overall picture. Students discuss their questions and observations with their teammates, and with practicing scientists, via the network. Although Kids Network attempted to transform students into scientists, several teachers felt the curriculum failed to provide authentic encounters with science; others felt that the science activities were unsophisticated intellectually and technologically.

The Contemporary Precalculus project developers did not explicitly claim that their course attempted to give students the experience of being applied mathematicians, yet much of the approach to problems incorporated in the materials they produced had that effect. Students worked with real data, developed their own mathematical models, and often had to write up reports as though they were consultants to a scientific enterprise or a corporation faced with a decision. Students who had taken the course performed relatively well on a national contest in mathematical modeling held each year and open to teams of high school and college students. One of the teachers at the North Carolina School of Science and Mathematics attributed this success not to specific knowledge that students had gained in the course but rather to the approach to realistic problems that they had learned. "It's not because they know more. It's because they can use what they know" (Kilpatrick et al. 1996, p. 163). Similarly, the Urban Mathematics Collaborative, by bringing students and teachers into contact with activities performed by mathematicians in business and industry, helped make the curriculum more authentic mathematically.

A Learner-Centered Pedagogy

Teachers of science and mathematics are increasingly being encouraged to formulate their instruction around the needs and interests of their students, to listen carefully to their questions and ideas, and to provide them with repeated opportunities for exploration and understanding. The child's view of the subject matter is seen as important because it can suggest to the teacher the direction that subsequent instruction might take.

Learner-centered pedagogies have been part of earlier reform agendas, dating back to the progressive school movement initiated by John Dewey and continuing through the school improvement efforts of the 1960s. Current forms of learner-centered pedagogies generally are based on constructivist philosophy, which views mathematical and scientific knowledge as socially constructed and mathematics and science learning as involving both individual knowledge construction and social enculturation. Constructivist teaching strategies, which incorporate the two approaches discussed above to a considerable extent, are intended to deal with certain phenomena observed in the classroom: that students bring their own ideas to the science or mathematics lesson, that these ideas come from students' "sense-making" of their experiences, and that some of these ideas (often dubbed naive conceptions or misconceptions) are difficult to dislodge and may reside side by side in a student's head with canonical scientific explanations. Solomon (1994, p. 7) points out that constructivism, as it has evolved, provides an explanation for these phenomena through combining and elaborating on three coexisting developments: a theory of personal constructs based on individual experiences; the notion of a "children's science" as contrasted to the science of teachers or the science of scientists; and research on the sociology of knowledge, including the consensual process by which scientific knowledge is built. Educators espousing a constructivist pedagogy would argue that "[T]he emphasis in learning is not on the correspondence with an external authority but the construction by the learner of schemes which are coherent and useful . . ." (Driver as quoted in Eisenhart and Marion 1996, p. 277). Thus, the emphasis in teaching is shifted from the teacher behaving as the authoritative voice of science or mathematics: "Learn this because it is correct"—to the students being guided by the teacher toward successful organization of their own developing ideas: "Does this make sense to you in view of your experiences?" Experiences here can be variously interpreted as students' everyday experiences outside the classroom and experiences explicitly staged in the classroom: well-designed practical activities, classroom discourse that mirrors scientific or mathematical discourse, apprentice-like inquiries that induct students into the practices of scientists (Driver et al. 1994).

Learner-centered pedagogies based on constructivist approaches can be identified in several of the cases we have studied, in which teachers are expected to deemphasize the view of mathematics and science enshrined in traditional textbooks and curriculum documents and, instead, to legitimate the thinking of the individual child as the starting point for instruction. The "child's science" and the "child's mathematics" are seen as not only valid in themselves but as essential for the teacher to understand and build upon. Discourse among students around topics being studied, with the teacher as guide, are part of constructivist teaching—the notion being that this type of instruction creates learning communities modeled on the construction of knowledge by scientific communities. Teachers involved in the California science education reform, for example, attempted to create a learner-centered pedagogy by mixing hands-on activities with constructivist teaching strategies. Constructivist views of learning were central to the development of the NCTM *Standards* even though these views are not elaborated explicitly in the document. Project 2061, although generally agnostic on teaching strategies, puts great emphasis on not only the acquisition of scientific knowledge but also the development of scientific habits of mind not unlike those of practicing scientists. On the other hand, the designers of Kids Network and Mimi did not explicitly hold to a constructivist conception of teaching or learning, and neither did the teachers who developed the Contemporary Precalculus course.

The academic community of scientists and mathematicians is not always aware of constructivist views and, when it is aware, is not always sympathetic. Although the term "constructivism" has become part of current educational jargon, science and mathematics educators as well are engaged in lively debate on the validity of the claims that undergird pedagogies based on constructivist philosophy and its most effective enactment in the science or mathematics classroom (see, for example, Bereiter 1994, Cobb 1994, Driver et al. 1994, Solomon 1994, von Glasersfeld 1995, and Phillips 1995). These educators generally agree, however, that the balance between the traditional teaching methods experienced by the large majority of students today and the more learner-centered methods advocated in several major science and mathematics reform efforts needs, in most classrooms, to be righted in favor of the latter.

Who Owns School Science and Mathematics in the 1990s?

As we noted at the outset of this chapter, the United States is moving out of a 40-year period in which university-based scientists and mathematicians have been the primary arbiters of the content to be taught in these fields to students in grades K-12. Of the eight innovations selected for this study, only Project 2061 has, as a matter of central principle, accorded the academic research community a key role in identifying the essential content to be taught at elementary and secondary school levels—though the project also consulted with scientists and mathematicians from industry and teachers in an extensive review process of its documents. The newly influential players in determining what is to be taught to today's students include, increasingly, the public at large, scientists outside academia, and teachers themselves.

Several factors have figured in this shift in curriculum power away from the academic community. For one, research scientists in universities no longer enjoy the influence they had in the decades immediately after World War II, not even in determining the scientific research agenda itself. Research costs money, and the public has many priorities for scarce public dollars. Elected officials who provide funding ask increasingly how research will have an impact on the country. Will it increase productivity? Will it help make a safer community? Will it contribute to a healthier society? There still is a place for research without apparent short-term payoffs, but it is less prominent on the public agenda, often over the objections of university-based scientists. Even at the National Science

Foundation—created, in part, to foster and protect basic research—priorities have shifted somewhat toward more applied work.

NSF is the key federal agency for improving science education and an important illustration of this trend. It was created right after World War II to advance fundamental science, which had proved so important in the Allied victory. Congress inserted a second mission into the founding legislation, however: to improve science education. In the early days of the agency, in fact, education expenditures at NSF totaled about 40 percent of the total budget, and were devoted mostly to the support of graduate students to ensure the continuing health of the scientific enterprise. It was not until after the Soviet Union launched Sputnik that NSF took on the responsibility for science education reform in elementary and secondary schools (Raizen 1991).

The link between university researchers and curriculum reform was to prove potent and long lasting—but not permanent. For example, NSF's systemic initiatives are designed to build partnerships among many government and private organizations for improving science and mathematics education, among which university faculty are only one type of partner. Our cases also reflect this trend: NSF now supports many projects for the schools in which university-based scientists no longer play a central role, a circumstance virtually unheard of during the last wave of science education reform in the 1960s. It is not that NSF is no longer the major federal agency for significant reform in science education. It is rather that the agency itself now has a more inclusive view of what is needed, reflecting deep changes in reformers' understanding of how schools change as well as changes in the broader scientific community itself. To illustrate the latter point, entirely apart from its activities in science education, NSF has moved toward according applied work higher priority and involving scientists outside the university to a greater degree than it did in the 1950s and 1960s. In fact, for much of the 1980s, NSF's director was an engineer from IBM.

As science itself moves toward more practical work, one justification for the trends toward applications that we see in the innovations is that the new curricula more accurately portray what science is like today. Furthermore, much scientific research, both basic and applied, is also interdisciplinary, which fact provides additional justification for the kind of subject integration that characterizes so many of the innovations we have reported. There are, however, additional influences that point in the same direction. The primary one in our studies has been the assertiveness of the teachers themselves and their involvement in curriculum design and development as discussed in the next chapter. They are claiming prerogatives in defining the *content* of school science and mathematics. This development is relatively new. It has always been assumed that teachers are the experts in devising pedagogical strategies, but that they have no particular expertise or authority when it comes to selecting content. This belief is being brought into question. In several of our cases, teachers assert that because they know the students best, they have an important role in choosing the topics that meet their needs as well as in selecting the instructional approaches they consider most effective. And as priorities shift from increasing the scientific manpower pool to ensuring scientific and mathematical literacy for all, teachers may legitimately claim greater understanding of elementary and secondary students' needs than is common among university faculty.

And so, as the ownership of science has become more diffuse in the late 20th century—and more contested—so has the ownership of science and mathematics education. More people are advancing claims for a legitimate voice in what should be taught. Scientists and mathematicians will always be the final arbiters of accuracy, but the possibilities for actual curriculum choices are vast, and the number of players is multiplying. If applications and subject integration remain prominent, we may be entering a period of even greater pluralism than we have seen in recent decades. There is a paradox here. The early nineties have been years marked by the development of national standards. Yet the actual curricula being designed gravitate more toward locally determined content in an attempt to make subjects relevant to student and community life. A coastal community in rural North Carolina faces different environmental challenges than does east Los Angeles. The curriculum is different, even if the general goals are not.

The clearest view of a likely curriculum future may reside in California. General standards (the *Framework* in California's case) have official standing and are accepted, but teachers are then accorded considerable latitude in designing the actual curriculum. Variation results, thus enhancing teacher commitment and energy—but all those involved believe that they are working toward the same goals.

4

The Changing Roles of Teachers

Norman L. Webb

Our eight case studies reveal a marked change in views of the importance of teachers to innovation. The contrast between teachers' current roles in innovation and their participation in developing and implementing innovation in the "New Math and Science" era of the 1960s is pronounced. Curriculum projects of the 1960s were not devoid of teacher participation, but clearly placed a greater emphasis on disciplinary rigor than on "teachability."

Within some of the cases, the place of teachers in the innovation even changed over time. ChemCom, initiated in 1981 by a grant to the American Chemical Society from the U.S. Department of Education, began with little teacher involvement, but over time it turned more to teachers for input, field testing, and dissemination. The Voyage of the Mimi, also initiated in 1981, had very little K-12 teacher involvement in its development. In contrast, California science education reform—the youngest innovation of the set, which was launched with the release of the 1990 *California Science Framework*—was heavily directed toward teachers and dependent on them.

Teachers' relationships to change and innovation are very intertwined with many other features. In order to gain any significant instructional change, one must reach vast numbers of teachers, who differ greatly in their beliefs, content knowledge, preferred teaching styles, and the conditions under which they teach. The differences among teachers are so great, and their participation so varied, that any classification of their roles greatly simplifies a multifarious reality. As Tolstoy cogently depicted in *War and Peace*, the great battles are won or lost by individuals in the field acting upon their own interests, motivation, and will to win, influenced little by the commanders, generals, and others acting from headquarters far behind the front lines.

Having teachers take prominent roles in development and implementation causes some dilemmas. Any large-scale educational innovation must be prescriptive enough to be used effectively by a large number of teachers, while flexible and general enough for individual teachers to mold to their specific needs and local requirements. Calling upon teachers to be primary developers of innovation increases the costs and slows the process. A question can be raised about how far teachers, as developers of innovation, can depart from the status quo, when their primary reality is greatly influenced by years spent teaching the traditional way. Some of the most challenging and interesting piano music has been composed by those who did not play the piano. However, ignoring the possibility that teachers can actively create innovation increases the likelihood that the innovation will be of no interest to teachers, ultimately minimizing its impact.

Our studies document a shift in strategic thinking about innovation and a change in the centrality of teachers to innovation. Three broad categories represent the variations in teacher roles from the past, and variations among different innovation efforts: teachers as designers of change, teachers as object of change, and teachers' changing roles in the classroom and toward their colleagues. Teachers, as parts of a system, do not have full autonomy over their professional work and daily classroom activities, nor do they expect to. What roles teachers do assume in initiating or advancing innovation are influenced by many other factors.

In this chapter, only teacher roles related to the three categories mentioned above will be discussed. This chapter attempts to isolate and magnify a few parts of the interactions between efforts at innovation and what teachers do. This is done in full awareness that any cell or particle that the lens brings into focus resides in a living organism. What is seen is influenced by innumerable other factors beyond view or less resolved, but of no less importance—factors such as diversity within the teaching force, rules, traditions, requirements of educational systems, and other countervailing forces that tend to keep things the way they are.

Traditional Roles of Teachers

Studying the changing roles of teachers implies considering the deviation from what teachers were doing before or what they were doing in some other context. Determining the extent of change implies the existence of some metric for representing the degree to which teachers' roles have deviated. We compared teachers' roles in the innovations we studied to traditional teacher roles as portrayed in the case studies, and to the generally held concept of what traditional teachers do. Through considering the goals and intent of the eight innovations, we gained some perspective on what teachers were doing that was seen as needing change.

The traditional teacher's role in the typical science or mathematics classroom is greatly governed by textbooks, with the teacher dominating conversation through lecturing and students passively listening or working individually on exercises. Teachers are to explain and show how. One teaching consultant for the California Science Implementation Network in the state's science reform strove to have teachers make their own connections among scientific ideas. She heard a common response from teachers, ". . . if I just had the book, it would provide the links" (Atkin et al. 1996c, p. 41). One of the North Carolina School of Science and Mathematics (NCSSM) teachers who wrote *Contemporary Precalculus Through Applications* reflected, "It used to be it was unfair of me to ask a student a question I hadn't taught them to solve. And I don't believe that any more . . . I think now my focus is much more [that] the way you do mathematics is you think about things" (Kilpatrick et al. 1996, p. 160).

A number of the cases illustrate a departure from the traditional roles of teachers by labeling a dominant force in specifying what is taught or what should be taught. Approaches to teaching and curriculum development are described as "teacher-driven," "textbook-driven," or "issues-driven." The Voyage of the Mimi materials, for example, are described as making a transition from a textbook-driven program to a multimedia approach. ChemCom's deviation from the norm is characterized by describing the new approach as an issues-driven curriculum. Two teachers, who were active in one of the urban mathematics collaboratives, described their pedagogical orientation prior to their participation in the collaborative—an orientation from which they were moving away—as teacher-centered, textbook-driven, and calling on students to learn algorithms and isolated skills. They were seeking a more student-centered approach to learning. Another teacher at a different collaborative site reflected on his teaching in the past, "Ten years ago, I might have still been doing only basic math and high school math . . . trying to get the kids to do their problems and the topic test and to learn to do fractions and division and lots of drill and practice sheets probably. I knew it wasn't doing much good. And then a few years later . . . I quit that" (Webb et al. 1996, p. 283).

Teacher Isolation

Isolation is another attribute of the traditional role of teachers that was earmarked for change in the innovations studied. A teacher at NCSSM who helped to develop Contemporary Precalculus found the interaction among members of the NCSSM mathematics department to be in stark contrast to common practice:

> In most schools, teachers work in their own little rooms. They do all these wonderful things maybe, but they rarely share them with anybody. The idea of sharing materials is really foreign to most teachers, especially in public schools. There are a lot of reasons for that. I don't really think it's selfishness. It's a matter of never having a common time they can meet. After teaching five or six classes, the last thing they want to do is have a meeting (Kilpatrick et al. 1996, p. 225).

Reducing teachers' sense of isolation from other teachers, recent developments in mathematics, and new applications of mathematics were major reasons that the Urban Mathematics Collaborative (UMC) project sought to develop collegiality among professional mathematicians and teachers. The goal for site participation in the UMC project was to offer teachers an escape from professional isolation by providing a forum for professional discourse among teachers and others. The intention was for teachers to become aware of ideas, resources, and opportunities, to have genuine access to them, and to have the authority to utilize them. Prior to the collaborative's inception, the opportunities, incentives, and purposes for sustained contact with other teachers reportedly were uncommon. One collaborative teacher, for example, referred to herself as "living in a cocoon" before she became active in a collaborative. In the California reform, teacher networks were incorporated as the major reform strategy, in part to keep teachers from operating in isolation. One intent of this approach was to build consensus among teachers who could not be effectively legislated into compliance.

In addition to the lack of interaction among teachers, isolation existed in traditional forms of teaching in other ways. A number of cases referred to traditional teaching as presenting isolated concepts and ideas in the absence of how these ideas were actually used. One teacher using ChemCom said, in reference to the advantage of the old form of teaching: "Personally, I find it relatively easy to develop curriculum and assessment tools for a subject taught in isolation and based on memorization of specific content or applications of concepts only within the confines of a test tube and an ideal, highly controlled laboratory setting" (Rowe et al. 1996, p. 567). The shift in instructional sequence depicted in the Precalculus case signified a similar reduction of content presented in isolation from context. The change was from a sequence of definition-theorem-proof-example-practice to situation-problem-data-model-solution. One teacher was attracted to Contemporary Precalculus because the program linked applications to modeling and functions, "Instead of telling students about the characteristics of various functions, we start out with data and talk about modeling the phenomena found in the data with functions" (Kilpatrick et al. 1996, p. 165).

Teacher Status

Most of the eight innovations studied were envisioned and initiated in the early 1980s, when the status of teachers was low. The rationale given for the need of collaboratives was the widespread concern about the teaching profession (Nelson 1994). Large teacher shortages were forecast in mathematics and sciences. Low numbers of college students were going into teacher training programs. Experienced mathematics and science teachers were leaving the profession for better paying computer-related jobs in industry. Those who became teachers were increasingly coming from the lower academic range of their graduating classes. Perhaps the lowest point in teachers' status was reached in 1983, when *A Nation at Risk* (National Commission on Excellence in Education 1983) attributed the low mathematics and science achievement among American students to unqualified teachers. Rates of reported burnout increased, as teachers felt more and more that they were locked into flat careers and working under very restrictive, less-than-desirable conditions.

Teachers in high schools and elementary schools lacked credibility even within their own content areas. High school teachers at NCSSM who developed and wrote a precalculus book were paid little attention by university mathematicians, perhaps because of the perception that high school teachers did not have a sound understanding of mathematics. The former chair of the mathematics department at NCSSM recalled skepticism about what high school teachers could do, ". . . I sensed that funding sources did not believe that high school teachers in North Carolina could do original work" (Kilpatrick et al. 1996, p. 206). This was the case even though several of these teachers had doctorates in the field. In Project 2061, tension existed between academic scientists and teachers over who was in the best position to select what science students needed to learn and could learn. A contributing reason given for launching Project 2061 was that teachers were poorly prepared. This lack of adequate preparation was given as a reason for many elementary school teachers not understanding even the most fundamental concepts in science and mathematics.

Teacher professional development was very narrowly conceived in the 1970s and 1980s. In part, this conception was an artifact deriving from the philosophy that all the essential information for teaching mathematics and science to students was contained between the covers of textbooks. Short, one- or two-session training, or an occasional full-day institute, were the prevalent forms of teacher in-services. The National Science Foundation (NSF) summer institutes associated with the 1960s mathematics and science reform era were relics of the past. A national survey conducted in 1985-86 found that more than half of the science and mathematics middle school and high school teachers reportedly had fewer than six hours of in-service training during the preceding 12 months (Weiss 1987). Little change in in-service hours was observed in a national survey conducted in 1993, which indicated that 44 to 45 percent of the mathematics and science teachers in grades 9-12 reported having fewer than 16 hours of in-service education in the preceding three years (Weiss 1994). During a time when advances in technology were rapidly creating new mathematics for solving computational algorithm problems, and expansions in genetics and biochemistry were opening new vistas in the world of science, inadequate support was given to keep teachers current.

Historical Significance of Teachers' Changing Roles

Significant shifts or evolution from past periods of reform also can be judged historically. They can be judged in relation to the magnitude of the problem being solved or addressed. Historically, the curriculum reforms of the 1960s centered on changing the content taught, broadly conceived as including both what to teach and how to teach it. Practicing teachers had a role in those reforms through field testing materials or, for a few, through serving on writing teams. The presence of K-12 teachers was overshadowed by the research mathematicians, scientists, and academicians. Subject matter content was the driving force. Encouraging the professional organization to exert more leadership and advice, in 1980 National Council of Teachers of Mathematics (NCTM) president Shirley Hill said in her inaugural address, "In the 1960s we learned that curriculum change is not a simple matter of devising, trying out, and proposing new programs. In the 1970s we learned that many pressures, from both inside and particularly outside the institution of the school, determine goals and directions and programs" (McLeod et al. 1996, p. 24).

The retrenchment of reform in the 1970s—the back-to-the-basics movement—said little about the role of teachers and focused more on what topics should be included in the curriculum. The rise of large-scale assessment and high-stakes testing in the 1980s took another tack toward change, by imposing minimum competencies for all students and requiring students to pass tests as partial requirements for graduation or grade promotion. Again, the imposition of testing requirements, adopted in some way by about a third of the states, said little about teaching practices and the role of teachers. These reforms were governed by the notion that the quality of education could be improved by mandating graduation or promotion requirements. This period in the rise of mandated testing was characterized by governments exercising what they saw as quality control means at their disposal while minimizing their costs. Testing a large group of students was less expensive than providing new materials or engaging in massive teacher professional development efforts.

The Diverse Teaching Force

Teachers differ in their knowledge of content, preferred style of teaching, experience, available resources, range of students being taught, age level of students, and conditions under which they teach. Many teachers across grades K-12 vary in their knowledge of mathematics and science, their teaching load, and their inclination toward engaging students in learning through activities. In reflecting on the changing roles of teachers as derived from our cases, "teachers" constitute a very diverse group. What is revealed in the studies are trends rather than precise characterizations.

The eight case studies of innovation identified and clearly revealed teachers who reported that they had changed their classroom practices and professional activities since becoming involved in the innovation. Many of the teachers attributed these changes to their participation in the given innovation. Some of the studies revealed teachers engaging in classroom practices not widely seen 10 or 15 years ago. Classroom observations revealed activities and practices, aligned with current reform visions, that were distinctly different from traditional norms. A middle school teacher who was active in the Memphis urban mathematics collaborative elaborated on what changes she had made: "In my fifth grade room you would have never seen manipulatives; you would have never seen group work, because I didn't do it five years ago. It was only after [I attended the collaborative's] Camp Mathagon that I started it and it's taken me awhile . . . "(Webb et al. 1996, p. 289).

But in considering all of the innovative projects studied, we find at least some targeted teachers who were unchanged and who continued to teach in the same traditional way they had taught for a number of years. Teachers are individuals and, as observed in the Voyage of the Mimi study, come with existing knowledge, persuasions, techniques, and views about the very constitution of mathematics and science. When an innovation confronts these strongly held beliefs, frequently it is the innovation that is changed. Innovations will undergo transformations, adaptations, resistance, and subterfuge because teachers vary in their knowledge of content and comfort with it, knowledge and mode of teaching, career longevity, years of curriculum use, and local demands and contexts. A teacher who was using the Voyage of the Mimi noted: "Someone else comes up with a new idea, a new way of trying out something or presenting it and we try it and see if it works. So, it seems to be somewhat of a fluid program, with us, too, constantly changing."

Clearly, this chapter is not on the changing role of *all* teachers. It is a discussion of teachers who have engaged in innovation and innovative projects and who have come out of this experience approaching their work and professional lives differently. It is a discussion of patterns of what teachers reported and were observed as doing, in more than one of these cases, that varied from the traditional perceptions of teaching. It is a discussion of how these eight cases depict teachers' participation in designing and implementing innovation in ways other than the norm.

Resisting Forces

Classroom teachers are members of departments in schools administered by school districts, which are governed by elected school boards and financed by local, state, and federal funds. Parents and the community place demands on the local schools and districts, which are already under a number of mandates from the state. Students who pass through the system go on to higher education, technical schools, the military, or the workforce, all of which look toward feeder schools to produce educated and well-prepared graduates. Within schools, teachers of primary-age children generally make decisions about student needs in as many as 12 content areas. Middle and secondary school teachers generally teach in one or two content areas for up to five or six classes, or a total of more than 100 students, in any one school day. And mathematics and science classes are not taught in isolation, but have to be coordinated with the scheduling of other content areas. The large and overwhelming forces beyond the control of teachers that govern the goals of education were recognized in 1980 by Shirley Hill, president of NCTM, as one reason why standards were needed. Hill encouraged NCTM to present advice on what the goals and objectives of mathematics education ought to be, because of the pressures from both inside and outside the schools that strongly influenced goals, directions, and programs.

New Roles for Teachers

Teachers do not operate in a vacuum, but work within a system of constraints. Even small shifts in what teachers do are significant in the context of the large, inert systems in which they work. Our collection of eight case studies reveals examples where innovation was advanced through teachers overcoming large resisting forces, as well as examples of attempted innovations that were too weak to nudge teachers into a slightly different orbit. This review of the eight studies does provide some illumination of the change process for teachers in their areas, including what efforts have led—or have not led—to teachers becoming more central to innovations, how teachers have changed their classroom practices, and what teachers have done to grow professionally.

Teachers' Changing Positions Within Innovations

Teachers' relationships to educational innovation are complex and multidimensional. The California reform was less an uprising by teachers for change as it was a synthesis of centralized and decentralized change. A synergism was created among a state framework, teacher networks, and other reform efforts where teachers were interacting with innovation in many different ways. Teachers helped shape the innovation while being a focus of innovation.

Teachers' relationships to innovation are complicated further by not being static. The goals, visions, and strategies for innovations vary over time. After five years, Project 2061 was still changing the ways it worked with teachers. The early plan changed over time. Educators in institutions of higher education used their academic knowledge to produce the vision in the document *Science for All Americans* (AAAS 1989). But when sites, which were selected in part based on the enthusiasm of teachers, were given the charge to have teachers use their craft knowledge to develop a model or models for translating the vision into classroom activities, some tensions arose between the sites and the central office. Some of the modules that resulted were seen by the home office as being too unrealistic and not based on how students learn. One member of the Washington staff explained: "A lot of [the modules] were good ideas, but they didn't grow out of having thought through how students would learn" (Atkin et al. 1996b, p. 220). Over time, adjustments were made in the strategies employed and in the scope of work.

Even in the innovations consisting mainly of the production of curriculum materials, where the teacher's role was primarily to implement the materials, over time teachers made their own adaptations to the materials. The Voyage of the Mimi materials, for example, were developed to be given to teachers. In trying to understand and use these materials, teachers employed a variety of strategies. Some teachers and administrators used a leadership model, in which teachers were trained to provide in-service training for other teachers. Other teachers collaborated with each other to understand better how the materials could be used with students and adapted as needed. Some of these teachers were selective in their use of the materials, bypassing the computer modules, centerpieces of the innovation. In this way, teachers shaped the innovation, more in spite of the intent of the innovators rather than because of it. Some of the ChemCom teachers were found, after two or three years, to have made adjustments in the curriculum by incorporating more mathematics or inserting a favorite topic. One ChemCom teacher explained why she taught the course differently from another teacher:

> The other teacher here and I are different in the way we approach things. We stay about the same as far as material is concerned, but we don't teach it exactly the same way. But that's fine, because he comes from more of a biology background and I have a physical science-type background but I have a strong interest in biochemistry; that's why I like teaching nutrition the best (Rowe et al. 1996, p. 553).

Innovations evolve over time, in part due to teachers' circumventing the canonical form of the innovation, but also due to the nature of innovative development itself. When an innovative idea was put into practice by teachers who differed greatly and who were in widely varying situations, mutations of the innovation often emerged.

Many terms have been used to describe the position of teachers in relation to innovation: top-down/bottom-up, grassroots, decentralized, classroom-based, teacher-driven, ownership, and collaboration with teachers, to name a few. How teachers were positioned with respect to the innovations described in the eight cases studied was equally diverse. Two dimensions help to depict the variations among the eight cases in the responsibility teachers had for the innovation, and in the emphasis given to changing teachers: teachers as designing and initiating the innovation; and teachers as the object of the innovation, or as what the innovation intended to change.

Teachers as Designers of Change

All eight innovations studied had a prime mover who assumed major responsibility for designing what the innovation should be. Sometimes, over the course of the innovation, the responsibility for sources of ideas shifted. In most situations, the people or the group that initiated the project assumed principal responsibility for specifying what the innovation would be. In some cases, teachers assumed a major responsibility in the design of innovations, a significant change from past practices. Reviews of 19 earlier NSF-funded science and mathematics materials development projects (Webb et al. 1993, Reynolds et al. 1993) found that none of the projects was directed by teachers.

The role of classroom teachers as designers of the eight innovations we studied varied, from teachers being the primary initiator and source of ideas for change to teachers having essentially no role, or only a minor role, in specifying what the change would be. In Table 4–1, the eight cases are positioned on a scale to indicate the roles teachers assumed as designers. Two of the science projects reviewed by Reynolds et al. (1993)—Kids Network and ChemCom—also were studied in our collection of case studies. In the summary of findings in the earlier study, Kids Network was identified as having no teacher involvement in the project development, whereas ChemCom was identified as having teacher involvement. These findings support the placement of these two projects on the designer scale as none and indirect, respectively.

Table 4-1.   Teachers as Designers of Change in Eight Innovations

Primary	Shared	Indirect	None
Primary initiator of change and principal source of ideas	Major responsibility for a specific part or input into shaping the change; valued for unique contribution	Provide significant feedback, reactions, or input to others who have major responsibility with some cooperative work	At most, used as a source of data by others in making decisions
Precalculus	UMC California reform	Project 2061 ChemCom NCTM Standards Voyage of the Mimi	Kids Network

The teachers at the North Carolina School of Science and Mathematics initiated the development of the Contemporary Precalculus course, in part, because of their dissatisfaction with students being taught abstract mathematical structures that had little relevance to students' current lives. Teachers in the mathematics department at NCSSM, sparked by the department chair, took it upon themselves to locate real-world problems that could be used to teach students mathematics and its applications. For teachers in a mathematics department to take this action was a significant departure from normal department activities.

In two of the cases, teachers did not initiate the innovations, but had major responsibility or participated equally in at least parts of them. The UMC project was based on the premise that empowered teachers would assume major responsibility for solving problems in educating inner-city youth. All of the collaboratives included teachers in major decisionmaking bodies. Teachers were given access to funding and other resources to initiate projects. In one collaborative, teacher councils assumed the major responsibility for a workshop series, and for providing professional development experiences for teachers in the designated schools. In California's reform, the role of teachers in the innovation changed as the innovation evolved. Toward the beginning, the California *Science Framework* was used to build consensus among teachers and others. Teacher networks, organized regionally, assumed responsibility for defining local curriculum changes in response to the *Framework*. An identified strength of the California reform was the empowerment of teachers to have ownership of the project and to buy into it. Teachers became designers of local change. The California reform acknowledged that teachers will implement a curriculum in many different ways and was structured to build on this reality. Staff developers in California reflected on ways they worked with school staff: "We create a balance between administration and classroom teacher. We help the classroom teacher feel free to plan and take action without the administration feeling threatened" (Atkin et al. 1996c, p. 35).

In four of the cases—Project 2061, ChemCom, NCTM Standards, and Voyage of the Mimi—even though teachers had input into the design or specification of what the change should be, their roles were restricted and limited by the structure of the innovation. In Project 2061, teachers came into the project after the vision for what students should know in science was set forth in *Science for All Americans* (AAAS 1989). When teachers became involved through the site programs, their roles were more to identify effective curricula for implementation than to define content. Teachers served as "essential helpers" and were viewed by the head office as consultants rather than primary decision-makers. Teachers in the local sites were insulated from the politics of the project by the site directors. Teacher input in developing models and benchmarks was valued and deemed essential to the long-range goals of the project, because the teachers provided a practical perspective different from the more theoretical perspectives of the academicians. The head office, however, maintained actual or perceived control; hence, we do not view the teachers in Project 2061 as having shared responsibility as designers of the innovation.

In the case of ChemCom, initially the American Chemical Society turned to chemistry professors to design units. After one year, talented high school teachers were put on teams to write units and to develop leaders to train other teachers. Teachers evolved into the prime movers for the curriculum by initiating, among other activities, support mechanisms such as an electronic bulletin board. In the NCTM (1989) *Curriculum and Evaluation Standards for School Mathematics*, 2 of 24 members of the writing team were classroom teachers. In this way, along with providing reactions to drafts of documents, teachers had input into what was included in the standards. The major writing, and the final forms of the standards, however, were the responsibilities of the team leaders and director, who were university and college professors.

The principal designers of the Voyage of the Mimi materials were a small group of writers and researchers nested in a college of education known for its emphasis on child development and social relevancy. Input was sought from teachers on the different components, but teachers mainly provided advice rather than made decisions. A teacher panel gave advice, while other teachers field tested materials. This multimedia program had numerous components that afforded a range of opportunities for teacher input. In the design of software materials, for example, teachers participated as curriculum advisors, field testers, and co-developers. The information from field testing that teachers produced was eye-opening for the developers in giving them knowledge about the meager technology in classrooms and other limiting school conditions. But even though a few teachers participated in some way in the design, the principal designers remained those in higher education. The set of materials was created in part to be so appealing and engaging to students that teachers would have to change in order to keep up with their students.

These four cases, constituting a wide spectrum of teacher responsibility, used teachers as indirect designers of change. We list them in Table 4-1 under one category, but in fact these cases varied considerably in the degree that teachers were engaged as designers of change.

The software and curriculum units used in Kids Network were developed by staff from the Technical Education Research Centers (TERC) and the National Geographic Society (NGS). Each of the seven units went through a developmental process that included field testing by teachers. Formative data were collected from teachers that were used to make adjustments in software, curriculum units, and procedures. Teachers produced the data, but TERC and NGS staff did everything else. Teachers had no responsibility for designing any of the materials.

Projects studied in earlier research included one or two teachers on the development team, but the primary involvement of teachers was in field testing and reacting to materials. Academicians and researchers from nonprofit research and development centers were the primary designers and developers. Within the collection of eight cases studied here, more than a third of the cases had significant teacher involvement in the design. This is a departure from how teachers previously were engaged, even in the recent past.

Teachers as Objects of Change

Another source of variation among the eight innovations, and between these eight and innovations in other reform periods, is the attention given to teacher change. In the 1960s, greater emphasis was given to developing curriculum materials.[1] Development of written materials was the dominant strategy for delivering new materials to massive numbers of students. NCTM leaders who helped initiate the development of the curriculum standards knew from their experience in the 1960s that just changing the curriculum was not sufficient to improve mathematics education. This knowledge, along with the perceived widespread shortage of

---

[1]Even though NSF invested far more in teacher institutes than in curriculum development, these institutes were aimed originally at upgrading teachers' subject matter and later on at implementing the new curricula.

qualified mathematics teachers in the early 1980s, strengthened their belief that any move away from traditional instruction would require strong programs for staff development and teacher preparation. As a consequence, along with curriculum standards, the organization developed teaching standards. NCTM's Instructional Issues Advisory Committee (IIAC) was given credit, in part, for developing the idea for standards. An IIAC member recalled the attention given to teaching from the early stages of thinking about standards: "Somehow we got onto the idea that what IIAC ought to do was define professional standards in general—not just for selection of textbook material but for curriculum, for teaching, and so on" (McLeod et al. 1996, p. 31).

A similar point was illustrated in the development of the Precalculus project materials. Not everything can be included in the written text. Teachers who did not receive any training in using the precalculus materials did not recognize all of the value in an activity. These teachers, with their students, frequently rushed through the book. In contrast, teachers who did have training were able to get their students to go much more deeply into the materials.

Over the past 30 years, increasing attention has been given to how content is taught, in addition to what content is taught. Considerable funding has been provided for teacher enhancement by NSF and other federal agencies (FCCSET 1993); a federal program (the Eisenhower program administered by the U.S. Department of Education) was instituted specifically to support teacher initiative and participation in professional development. Teachers are spending more time in professional development activities, and teacher membership in professional organizations has increased. Changing what teachers do in the classroom, beyond just changing what is being presented, has become more important to innovation.

As objects of change, teachers are considered to be very important to the change process. Within an innovation where teachers are objects of change, it is their knowledge of the content area, their pedagogical practices, and their responsibilities that become the focus of the innovation. Viewing teachers as objects of change reflects not only the intent or purpose of the innovation, but the strategic approach to how change is advanced.

Table 4-2 depicts the placement of the eight innovations on a scale representing, to a greater or lesser degree, how teachers were viewed and treated as objects of change by the innovation. The UMC project was created on the assumption that empowered teachers would develop the solutions needed to improve education in inner-city schools. Reducing teacher isolation—from other teachers, as well as from recent changes in the field of mathematics and its applications, from research mathematicians, and from users of mathematics— was felt to be critical for teachers to find the curriculum materials, new approaches to teaching, and support they would need to make significant improvements. Teachers were the primary objects of change. The scale of change needed in the California reform forced an approach for implementation that would depend heavily on teachers. Three critical components of change characterized the reform effort: building a consensus on goals; building an implementation structure; and creating incentives for educators to find their own path to the shared goals. Teacher professional development networks were thought essential to help persuade teachers of the new vision for science education. The California effort counted heavily on teachers, along with other components of the system, to create the changes necessary for reforming science education. The main vehicle of the innovation by the NCSSM teachers was institutes for teachers supported by the textbook. Professional development institutes were deemed important for teachers to fully understand the richness in the materials. The printed materials alone could not communicate fully the approach to learning.

Table 4-2.   Teachers as Objects of Change in Eight Innovations

Primary	Critical	Interactive	Minor	None
Other changes depend on teachers initiating change in their knowledge, practices, and leadership	Teachers are essential to building consensus for change	Teaching and what teachers do interact with change in content and institution, so that all have to be changed at once	Materials specify what is taught and how it is taught; teachers change in executing the materials	Problem is not teachers or teaching, but content and what is taught or imposed learning requirements
UMC	California reform	NCTM standards	Project 2061 ChemCom Voyage of the Mimi Kids Network	
	Precalculus			

The NCTM standards were based on a more systemic approach to change, which acknowledged that teachers and their teaching both needed to change along with curriculum and institutional change. In both the NCTM standards and the Precalculus innovations, teachers were targeted for change along with the content.

The four other innovations gave some significance to changing teachers, but the dominant strategy for change was to vary the content that was taught and the materials that were used in the teaching of that content. These four innovations were not as notably different from those of the 1960s with respect to how teachers were viewed as to what needed changing. After Project 2061 produced the guiding document, *Science for All Americans*, through the process of classroom teachers developing implementation models, a second document evolved, *Benchmarks for Science Literacy* (AAAS 1993). The 855 benchmarks are statements of what science students should know at specific grade levels. Neither document is committed to any particular instructional strategy. The focus is more on what students should know rather than on how students should be taught. ChemCom is a chemistry program for high school students that broke the tradition in its goals, organization, and choice of content. A textbook for a year-long course evolved from special topic modules. Teachers served together with university chemists or chemistry educators on development teams. Even though teachers were expected to change their teaching strategies, the role of teachers with respect to shaping the innovation itself evolved into dissemination and training. Trained teachers were important to the innovation, but the main focus of change was on content—having a chemistry course organized around issues and including in the content discussion and decisionmaking on the issues.

The first *Voyage of the Mimi* and the *Second Voyage of the Mimi* are multimedia packages designed to supplement the standard curriculum. The video story was designed to "hook" children and make them curious. "What teachers did with it might be much more rigorous and sequentially organized . . . might look much more like conventional science," explained the originator and executive director (Middlebrooks et al. 1996, p. 413). Some teacher training was associated with the program, but it was never central. The major changes were built into the materials. Kids Network is a modular curriculum for grades 4-6 of seven six-week units each online several times a year. All units include a teacher's guide and consist of a series of cooperative science experiments in which students use the telecommunications network, which is part of the innovation, to share their data with other sites. The materials were designed to shift from more conventional practices to a hands-on, group-centered, and dynamic approach to science teaching. Teachers could execute the units by being true to the structured hands-on curriculum. As with the Mimi materials, the approach to teaching was built into the curriculum.

Teachers' Changing Role in the Classroom

The difficulties that teachers face in incorporating innovative programs into their classrooms have been documented as overwhelming (Fullan 1991, Huberman and Miles 1984). Teachers are faced with day-to-day coping, unsuccessful attempts, successive cycles of trial and error, and the sacrifice of other core activities. Difficulties arise from an overload of having to do simultaneous tasks, not knowing if what is being tried will work, putting into practice a program without full understanding of what that program is, and developing a personal interpretation of what is new and what has been considered appropriate for a number of years. Even when teachers receive training on new curricula or teaching methods, it is common for them to incorporate some of the new curriculum ideas, but to adapt them to fit their traditional teaching behavior (Porter 1989, Stodolsky 1988).

Changing Classroom Practice. None of the eight innovations reached a magnitude of change in classroom practices that could be characterized as uniformly applied across a significant proportion of science and mathematics teachers in a state or the nation. For change of this magnitude to be concentrated and powerful enough to overcome the system's inertia would require far more time and resources than any one of the eight innovations had at its disposal. What the innovations were able to do was provide a number of teachers with an awareness of what change is needed, and provide some individual teachers with sufficient motivation and support for noticeable shifts in classroom practices. The case study report indicates that the NCTM *Standards* have at least been studied by about three out of every five secondary mathematics teachers in the country. And there is evidence that at least some teachers are thinking differently about mathematics and science, and are engaging their students to learn in new ways. But the number of these teachers is relatively small compared to the number in a state, or the nation, who could be reached.

Seeking evidence of changes in classroom practice, the case study report of the NCTM *Standards* notes that "change comes slowly, or not at all, for many teachers, even those in reasonably progressive school settings" (McLeod et al. 1996, p. 88). Certain reasons were revealed in the study as to why this was so. Teachers were caught in the middle between supervisors influenced by the standards, who wanted more problem-oriented teaching, and parents who wanted their children to be good at basic mathematics. In this context of feeling pressured by conflicting aims, teachers cautiously moved toward using story problems. Their rate of change was tempered by the need to demonstrate how groups of children involved in mathematical investigations could solve real and complex problems. Teachers are comfortable with highly repetitive teaching based on test results showing that "the children still haven't mastered it." Very strong efforts are required on their part to become comfortable with their ability to discern whether students are advancing toward conceptualization and understanding.

Amidst active reform efforts, greater demands are placed on teachers to deal with ambiguity. A school district in transition can have students from the elementary grades enter middle grades with preparation that is different from that of previous years; their knowledge could be misrepresented when a traditional test is used to place these students. In the throes of reform, teachers are forced to confront different philosophies. A teacher who believes that students are to be picked up from where they are and advanced as far as possible has difficulty working with teachers who believe in a highly structured approach to learning, where topics have to be taught in a specific sequence.

Adaptation. Teachers need to understand the innovation. From the study of the development and implementation of Voyage of the Mimi, it became apparent that teachers could implement an innovation without teaching innovatively. The assumption of the developers was that some teachers would probably teach "adventurously," but little direction was given in the materials about what teachers should do in interacting with students. This was done intentionally, so that the materials could be used in a variety of classrooms and with a variety of teaching styles. The director expressed the expectation that teachers would use the materials in any combination, conventionally or adventurously. The storyline in the video and the other materials were designed to hook students, who then would bring the teachers along. This approach was not very effective in getting those teachers to change who had strongly held beliefs about the makeup of school mathematics and science. Researchers concluded, "science instruction continued to be short, mostly talk, and teacher directed" (Middlebrooks et al. 1996, p. 505). Teachers were found to use the teacher's guide more prescriptively, rather than be adventurous. One teacher explained, "Basically, [Mimi is] a descriptive program: this is what you do. You can change it around and modify it, but it pretty well follows procedures already established" (Middlebrooks et al. 1996, p. 447).

Teachers modified and molded innovative materials to better fit their own conceptions of what students should learn, how they should learn, and how much effort should be expended to make change. ChemCom teachers needed content that appealed to a more diverse student population. As a result, more students with mathematics deficiencies enrolled in the ChemCom course. This forced some teachers to modify the course based on what the students could do mathematically. One teacher commented:

> We've tried to open up this course to all kinds of students and found out that certain kids still cannot get the math that we utilize in ChemCom. I found their grades went down a lot this year because we're bringing in a new kind of student and we thought they could handle it, but their algebra skills, if they had them, were very poor. We found out that they had a hard time mastering the math we had. So we modified the ChemCom course around that and even dropped the math level a little lower than they had in the book (Rowe et al. 1996, p. 549).

The lowered mathematics requirements did not always remain lowered. After two or three years of teaching ChemCom, teachers were likely to adapt the course by including more mathematics. The mere length of the course forced teachers to become selective of content. Some evidence was provided in the case study that the willingness of teachers to meddle with the content of the course depended on the teacher's chemistry background. Chemistry specialists were more vocal about important content omissions. Nonchemistry majors were more likely to retain the specified content.

Professional Growth. Seven elements of professional growth that encapsulate teacher change emerged from the analysis of the UMC project. The importance of these elements is reinforced by findings from the other seven studies on the difficulty of achieving significant change by teachers in their classrooms and the amount of effort needed. The following seven elements were derived from the existing research and from observations in studying the UMC project:

- **Disequilibrium.** Teachers have to have a reason to mount the required effort to change. There has to be some dissatisfaction with their current practices and an understanding that the new approach would be better.

- **Exposure.** Teachers are not inherently opposed to new practices, but they are not always aware that there are alternatives to what they are doing. Hearing and learning about new programs and methods of teaching are important to getting teachers to buy into change.

- **Existence proof.** Being given evidence by another teacher that a new program works is very powerful.

- **Modeling.** Hearing about an innovation, or being given an innovative program, is generally insufficient for a teacher to implement the program. Professional development experiences where teachers can see how other teachers used the new materials, or can observe other teachers in their classrooms using new approaches, are important vehicles for helping teachers make a transition.

- **Support.** Being innovative and implementing an innovation are difficult and take time. Whatever the program, eventually teachers have to do it themselves. As teachers customize the innovation to make it work within their classrooms and with their students, having someone to confer with about questions, intercede parents who object, help arrange the necessary resources, and collect the necessary equipment is very important.

- **Experimentation.** Customization and risktaking are critical to most innovations. Because of the many variables that exist in a teaching situation, boilerplate programs do not exist. Teachers need a willingness and opportunity to try new ideas and to find out what can work within their individual situations.

- **Reflection.** This element is very closely coupled with experimentation. Experimentation is of little value if teachers do not have the time to think about what they did, how it worked, and what could be done differently.

These elements leading toward professional growth of teachers do not constitute a formula or guarantee of significant change, but they were evident in the set of studies associated with what successful teachers had experienced in making change happen.

Across the eight innovations, teachers reported—and were observed— teaching in new ways attributed to their participation in the project. Some teachers had changed from a more teacher-centered classroom toward a more student-centered classroom. Other teachers assumed more of a role as a facilitator of learning, enhancing the classroom discourse, expanding the variety of methods for assessing, drawing more on practical applications, and providing students with more opportunity to integrate and connect scientific and mathematical ideas through organizing content by themes and big ideas. An engineer, who had worked with a teacher (Jo) as a summer intern in association with the Milwaukee urban mathematics collaborative, reported that Jo had infused blueprinting into her geometry class. As he described:

> Jo has had a couple of units since she's been here on . . . blueprinting . . . in the math program that she puts together. Instead of just doing, "This is geometry. Look, there's a square." And everybody goes, "Great, that's a square." They sketch up a part . . . And then they draw it up as if it were going to be made as a blueprint [using] . . . two-dimensional drafting views (Webb et al. 1996, p. 291).

Teachers' Changing Role as Colleagues

A noticeable degree of activity among teachers was apparent in most of the eight innovations. The UMC project and the California reform addressed, in part, teacher isolation. Nearly 20 years ago, in a national survey of mathematics and science teachers, Weiss (1978) indicated that teachers valued collegiality over direct instruction from a supervisor. A significant effort had been made in several of the eight cases to bring teachers together so that they could support each other and provide each other with information.

Networking and Collaboration. In California, state leaders turned away from traditional implementation strategies of weekend workshops and one-shot in-services to strive for systemic reform in science education. The centerpiece of that state's change strategy became two state-funded teacher networks, the California Science Implementation Network and the Scope, Sequence, and Coordination Network. Here, a network represents an administrative structure devoted to bringing teachers together and fostering their communication with each other, both formally in institutes and informally between individual teachers. These networks served as the primary means to inform and support science teachers as they worked toward changes reflected in the state's science framework.

Both the UMC project and the California reform developed their strategies based on the growing literature on the value of teachers working together. The teachers at NCSSM who developed the Precalculus program bonded together because of the new roles that came from developing an innovative course. These teachers assumed the roles of "learner of mathematics, teacher of mathematics to peers, self-reflective teacher, and pedagogical experimenter" (Kilpatrick et al. 1996, p. 203). Other communities of teachers developed locally, through a common desire to change their teaching and their need for mutual support. These communities formed within mathematics departments among those teachers who used the Contemporary Precalculus materials, and among teachers who attended institutes conducted by the NCSSM teachers.

Leadership. Along with bringing teachers together through networking and collaboration, another common activity among several of the innovations studied was the development of teachers as leaders. In some cases leadership in implementing the reform was thrust upon the teachers. The California Science Implementation Network was based on the premise of shared leadership: It was neither top-down nor bottom-up. Underlying this premise was the strong belief that educational change can only be achieved by building leadership capacity both locally at the school site and statewide. This project was building at a time that many districts were eliminating science coordinators' positions, along with those in other content areas, to reduce costs. This trend forced teachers to emerge as leaders if there was to be any leadership at all.

One function teachers had to assume was making a case for reform. Teachers involved in the California reform had to justify to administrators the need for additional resources for implementing a new program. At the state level, 12 teachers with extensive classroom experience and with expertise in staff development and science content were assigned as teaching consultants, responsible for specific districts or regions. These consultants served to support teachers and connect them to the broader context of science education reform.

Developing leadership among teachers was addressed in different ways by the eight innovations. Some of the projects, such as the California reform, the UMC project, and ChemCom, provided specific training. Leadership training institutes were conducted for a few UMC teachers to help them understand more about planning for change, garnering the necessary resources to support change, and how to work with those who have the power to make change. The American Chemical Society sponsored two-week workshops to train teacher leaders for ChemCom. Teachers were given training in how to organize, run, and evaluate training workshops. Trained teachers, leading workshops using American Chemical Society guidelines, were deemed effective because a specified agenda was implemented by skilled practitioners who made the claims of the program believable to other teachers and who could recommend pedagogical strategies with credibility.

In implementing the Voyage of the Mimi materials, teachers assumed leadership roles locally out of need. There is some reason to believe that the program would have collapsed in most places without the network of mutual assistance that developed. In several sites, one teacher became the local technology specialist and served as resource, trainer, and cheerleader for the others using Mimi's software components.

Professional Development. Findings from the eight case studies depict a changing notion of professional development. Traditionally, professional development is viewed as a workshop or institute teachers attend to learn specific methods for implementing curriculum or practice. In the eight innovations, teachers engaged in a broad range of professional development experiences to learn about, and become proficient in, implementing change. Networks, communities of teachers, and leadership roles contributed to this expanding notion of professional development. One of the leaders of California science education reform described the changing conception of professional development as a shift in thinking, moving from what is "going to make me better for Monday morning" to understanding that significant change will take time and commitment. A staff member of the National Geographic Society emphasized the difficulty of getting teachers using Kids Network to reflect on pedagogy by considering how to help students ask better questions, how to make the electronic plotting of teammates' data more meaningful, and how to afford students with more opportunities to do experiments. The staff member defined the problem as: "How do you train a teacher, who's taught for 15 years from a lecture method, all of a sudden to quit lecturing, and to put the skills of learning into the hands of the children? That's hard . . . " (Karlan et al. 1996, p. 346). She emphasized the necessity of providing teachers with ongoing assistance from someone within a school system in order to help them through this transition.

Change in the conception of professional development depicted in these studies is due, in part, to the changing roles that teachers have assumed as professionals and, in part, to the nature of the innovations themselves. One cause of this expanded view of professional development is that the innovations themselves are directed toward teachers implementing a curriculum focusing on student inquiry, problem solving, reasoning, and applications. Students using ChemCom or the Precalculus materials were more apt to raise questions, seek solutions, and make assumptions in ways that were very difficult to anticipate. This view of instruction is very different from what many teachers have done over a number of years. In the NCTM case study, one consultant reportedly observed that teachers need to resolve what is important to them. This consultant remarked: "The magnitude of the change to an interactive and interpretive problem-solving pedagogy was so great that each minute of staff development time was needed for that" (McLeod et al. 1996, p. 95).

Another reason for the changing conception of professional development, as reflected in the cases, is that classroom teachers are being valued as specialists with expertise that others in the educational enterprise do not have. Clearly, those who study and write about education and those who have left the classroom have knowledge and experiences. However, these people do not have the same credibility as teachers—who themselves are under the same immense pressures as their colleagues—in explaining and describing how teachers can work with students.

Teachers are very concerned about doing the best they can for their students. They realize that, in some cases, their biology course may be the only biology a student will ever take. They are concerned that, if their students are to perform well in the advanced algebra class, they will have to be accomplished in some of the basic algebraic ideas, which are taught only in their introductory course. Because of this deep concern for their students' welfare, teachers are cautious about making radical changes without knowing that the change will be in their students' best interests. Hearing this message from a teacher who has successfully used the innovative materials can be very meaningful to those who are trying to be convinced. But even with training and ongoing support from other teachers, examples of teachers abound in the case study reports who incorporated or supplemented the new program with drill and practice or a favorite unit, if they felt these old materials had served students well in the past.

Within schools, the dynamics among teachers in a science or mathematics departments are changing. Teachers are becoming less isolated, through the initiative of teachers and through the support given by administrators. Rather than being confined strictly to their classrooms, teachers are finding ways to engage in cooperative planning, experimentation, and reflection. Working with colleagues in these ways has expanded the comfort zone for teachers, increasing their willingness to try new ideas or take some risks. The value of collegial support is demonstrated in some of the cases by their inclusion of hard-to-reach or hard-to-convince teachers, and innovation strategies that sustain efforts long enough for teachers to make the difficult transition required to institute significant change.

Teachers understand the complexity of change and the need to customize an innovation to their situations. Teachers have become consumers of professional development. As consumers, they are selective, find ongoing support, and seek what information they need to achieve the innovation they believe in. In this respect, teachers have become empowered.

Conclusion

Our eight case studies reveal that change has occurred more through the efforts of individual teachers than through efforts by departments, schools, districts, or states. Even though some of the innovations targeted departments of teachers as the unit of change, change was infrequently apparent across teachers within the departments, the Precalculus course at NCSSM being an exception. Clusters of teachers have marshaled innovation within their groups, but these clusters generally are formed by teachers linked in ways other than administratively. Teachers from different schools—and even from different districts—have formed bonds through attending the same institute, forming a network, serving on a task force together, writing a curriculum, or piloting a program. These teachers then turn to each other through the Internet, at professional meetings, and over the telephone to provide support to each other and to further the process of change. In some cases, the links in innovation formed among participants are stronger than the links among teachers within the same department or school who may be teaching other courses, be at different places in their careers, have different interests, or have different personal demands.

Change has come from teachers who have volunteered to participate. Mandated change has not been a critical feature of any of the innovation designs, nor has it been evident in any of the case studies themselves. Teachers who participated in the eight innovations had some professional autonomy and a strong interest to do the best for their students. Teacher participation in the innovations was motivated by this interest in their students. The participating teachers were "convincible." The innovations, however, were less successful with or did not have the time or resources to entice hard-to-reach or disinterested teachers. The teachers who had been reached by the eight innovations studied provide powerful stories of what could be. What is less clear from the collection of studies is how to reach that critical mass of teachers within a school, district, or state who can tilt the scale toward large systemic reform.

5

The Changing Conceptions of Reform

Senta A. Raizen
Douglas B. McLeod
Mary Budd Rowe

In this chapter, we examine some assumptions made in the eight innovations we studied on how to bring about educational change. These assumptions—sometimes explicit, and sometimes traceable through a project's dissemination and implementation strategies—grew out of the general context for education reform briefly discussed in chapter 2: research on educational change, changing approaches to reform, and lessons learned from the 1960s. This chapter begins by briefly reviewing the evolving knowledge on how change comes about in education and the lessons of the 1960s. We then discuss some of the changing approaches to reform that have characterized the last decade: systemic efforts simultaneously addressing several critical educational elements, setting educational goals through the formulation of standards, and widespread involvement through partnerships and collaboration—and how these approaches are reflected in the NCTM standards project, Project 2061, the California science education reform, and the Urban Mathematics Collaborative (UMC). Last, we illustrate new approaches to dissemination and implementation as found in both the comprehensive projects and the curriculum development projects we studied.

Lessons From Research and Experience

Over the last several decades, there has evolved a considerable body of research on how educational systems change and individual educators take up innovations: the adaptations that reform initiatives meet in the classroom, the factors that favor or inhibit change, what makes for the success or failure of an innovation, and how success can finally be defined as institutionalization of the innovation within the system (Yin 1978).

Models of Educational Change

Conceptions of the educational change process are still in a state of flux. As research knowledge has accumulated, views have been changing from the linear to the diffusion model, and from that to the more active problem-solving model. As Hutchinson and Huberman (1993) point out, the changes in the research perspective on the process of reform in education run parallel to the changes in perspective on learning and teaching.[1] For example, the individual who is expected to adopt an educational innovation is now viewed as an active problem solver and constructor of knowledge, rather than as one who absorbs information passively.

Most of the eight projects we studied did not deliberately set out to adopt any particular model of educational change. Nevertheless, the influence of one or another—and sometimes a combination—of the models can quite readily be traced.

The Linear Model. The early linear model generally credited to Guba (1968) and Havelock (1969) saw planned change in education as based on a cycle of research, development, diffusion, and evaluation (RDD&E model). This influential work helped spur the U.S. Department of Education to establish educational research centers to create research knowledge and Regional Laboratories to diffuse that knowledge to the educational systems within the regions. The model soon drew criticism for envisioning a one-way flow from basic knowledge to user application—an approach that proved too simplistic and incapable of separating the different contexts faced by a given research-based change, the costs to teachers and others of making the change, and the essentially conservative nature of education (Berman and McLaughlin 1978). Nevertheless, this model still undergirds many innovations, including the Voyage of the Mimi, Kids Network, and ChemCom, all of which followed the path from careful development to dissemination and adoption.

A factor that complicates the adoption and implementation of such innovations is the often complex nature of the evidence on its benefits. A curriculum may prove effective while under the control of its developers, but not demonstrate the same degree of effectiveness once it experiences second- and third-generation implementation. Much research over the last two decades on the introduction of reform initiatives in education (e.g., Fullan 1993, McLaughlin 1990) has documented the inevitable change in any innovation made by adopting practitioners. This is clearly evidenced in several of our cases as detailed in chapter 7, particularly in those like Mimi, ChemCom, and Kids Network, which had a sufficient history of classroom implementation at the time of our research to illuminate the process of local adaptation. Interestingly, Project 2061, the reform initiative that built adaptation into its very design by selecting local districts to create alternative curriculum models, found the initial efforts at local adaptation wanting with respect to the vision of science education set out in its guiding document, *Science for All Americans*.

---

[1]Hutchinson and Huberman reviewed in detail the research on knowledge use and dissemination in mathematics and science education.

The Social Intervention Model. An alternative model, drawn from research on the agriculture research and development (R&D) system (Rogers 1962 and 1988) and the diffusion of new medical treatments took into account the social interactions among the individuals involved in the RDD&E process. Based on findings from rural sociology, this model posited the notions that face-to-face interaction was critical to the adoption of new practices, and that disseminators introducing innovations needed to be near in social status and interest to prospective adopters. The reliance on agricultural extension agents charged with linking university research to practicing farmers seemed to explain the success of agricultural R&D system in introducing farmers to demonstrated improvements in agricultural practices and to their successful implementation. The National Diffusion Network, created to disseminate validated improvements in educational practice, was directly based on the conception of documenting the effects of an innovation and then using linking agents drawn from the education rather than the research/development community to disseminate it and help teachers with adoption and implementation. The model continues to be influential, as evidenced by the establishment of the Eisenhower National Clearinghouse and associated Regional Consortia to inform teachers and other educators of the wide array of science and mathematics materials available, help them select those appropriate to their goals, and provide guidance for their effective use.

Although the "linker" principle is now well-accepted, a number of other factors come into play in the implementation of innovative practice. For example, it is considerably easier to introduce a new drug successfully, which requires no change of behavior in doctors, than to introduce a new surgical procedure for which special training is required, or even to abolish an ineffective one that has established itself in practice. This simply illustrates the now-recognized truism that the less behavior change a particular innovation requires, the more likely its adoption with a fair degree of fidelity, but also the less ambitious the scope of the adoption. Such fidelity usually means that little is being attempted (Huberman and Miles 1984). In fact, Fullan (1993, p. 26) claims that "the absence of problems is usually a sign that not much is being attempted. Smoothness in the early stages of a change effort is a sure sign that superficial or trivial change is being substituted for substantial change attempts."

All of the U.S. innovations encountered bumps in the road because they represent major restructuring and change in what is to be learned and how it is to be taught. They are bold ventures in terms of what they are trying to undertake. Even the four innovations that are more limited, tackling a specific course or content sequence, require considerable change on the part of teachers—and sometimes of schools—to be implemented fully.

The Problem–Solving or Constructivist Model. Such considerations have led to a third, deliberately nondirective, model of educational change: the user as problemsolver. This has an obvious counterpart in the new conception of the student in school as an active constructor of scientific and mathematical meaning rather than a passive recipient of knowledge constructed by others and handed down by the teacher. The Urban Mathematics Collaborative is a direct application of this model (Romberg and Pitman 1994), but the model also underlies the emphasis on teacher networks as agents for interpreting and implementing the California science education reform. One issue plaguing this model, as illustrated by both cases, is a source for new ideas to generate productive approaches to the problem(s) to be solved. As it happened, the National Council of Teachers of Mathematics (NCTM) *Standards* came along as a resource for the teachers involved in the collaborative; the state's *Science Framework* served as an anchor for the California teachers. Eventually, however, schools were asked to develop content matrices [to specify the science to be taught locally]. Perhaps the most successful example of this model is the Precalculus project, where teachers had both the motivation and the subject matter knowledge to address the problem they had set for themselves of teaching their course through applications rather than through abstract mathematical structures. Project 2061 in a sense is a combination of the first and third models: The project presumed that the six school districts enlisted to develop curriculum models based on *Science for All Americans* would successfully translate the scientists' visions into classroom reality; the process was fraught with difficulties.

Linking the "What" to the "How." There is an overriding problem with many reform attempts in education: They tend to be generic, that is, not linked to any particular subject matter content. Ignoring content—the "what" of education—and dealing only with the "how"—decentralizing governance authority, changing funding allocations, restructuring schools, empowering teachers, introducing educational technology—may derail reform and generate inappropriate research and misleading conclusions on the impact of reform. Just as the understanding of teaching and learning must be done in context, including the content of what is to be learned, so must educational improvement be considered in context, including subject-matter-specific learning goals. Our case studies advance this perspective in that they allow an analysis of major reform trends in educational improvement through the lens of specific content areas.

Lessons From the 1960s

Raizen (1991) discusses some of the differences in current educational reform efforts compared to a generation ago. Notably different is the involvement of the states (which was largely absent in the earlier reform wave); this includes mandates for more rigorous high school graduation requirements, guidelines for course content through state frameworks, and emphasis on accountability systems through increased student testing. Current efforts seek private sector involvement as well, specifically with textbook publishers, but also partnerships with industry in general. The Kids Network provides an example. As one of the original "Triad" projects funded by the National Science Foundation (NSF) in 1987, it required collaboration between curriculum developer, materials distributor, and schools; the roles played by private foundations as well local industry were noted earlier (see chapter 2).

These more inclusive approaches represent in part a reaction to the innovations of the 1960s, perceived in retrospect as a collection of isolated projects. This perception, however, is not quite accurate. At least in their later phases, these reforms reflected a concern with implementation of federally sponsored innovations, and so an emphasis within NSF on integration of teacher training with development of innovative course content. According to one source (Welch 1979), by 1975, 80 percent of NSF funding for training teachers was devoted to learning to implement new curricula. The U.S. Department of Education, for its part, created several mechanisms for helping schools institute promising curricula and teaching reforms. Among these mechanisms was the aforementioned National Diffusion Network, devoted to spreading effective curricula, and the system of Regional Education Laboratories that were to bring innovations to schools.

Systemic Reform. Despite these dissemination initiatives, when concern with the adequacy of mathematics and science education reemerged in the 1980s, the earlier reform efforts were deemed inadequate, if not outright failures. As noted in chapter 2, policymakers sought new approaches based on the accumulating research about change in education systems (and other organizations) and on the lessons learned from the 1960s. One of the most visible of these lessons was the recognition that individual projects, no matter how effective in themselves, no longer sufficed to meet the ambitious goal of scientific and mathematical literacy for all. Thus, the first lesson learned was that all aspects of the education system must be tackled simultaneously, an approach that has been characterized "systemic" (Smith and O'Day 1991). The objective was to make a comprehensive set of changes work together, not just teacher institutes in isolation, course content improvement in isolation, or student enrichment programs in isolation. The emphasis was on the education system at the state or district level, even though change still had to happen classroom by classroom.

The series of programs initiated by NSF in 1991 with the State Systemic Initiatives is probably the most visible example of this approach. NSF guidelines for systemic initiatives generally call for alignment of policy and practice with demanding goals for student learning, with particular attention to developing competent teachers (through both in-service and preservice training), challenging curriculum content and up-to-date teaching practice, assessment of student learning in line with goals and curricular content, and appropriate uses of educational technology. Long-standing programs—for example, those in support of teacher enhancement—are now framing their guidelines in the context of systemic reform (NSF 1994a). Active partnerships with stakeholders inside and outside education are seen as critical to any systemic improvement effort. An integral part of systemic reform is the formulation of a vision for education, often in the form of standards for a given school subject such as mathematics or science.

Standards. The call for standards, too, is partly rooted in the experiences of the 1960s. A second lesson from that era was the seeming ineffectiveness of letting schools and districts determine for themselves what they were going to teach and what represented adequate student achievement. This lesson appeared to emerge from the experience of the states, which were dissatisfied with the caliber of their students after a decade of trying to enforce a minimum of learning through specifying minimum competencies or using standardized testing. As described earlier, the lesson was reenforced by the better performance on international science and mathematics tests of students from centralized education systems such as Japan and Korea. In consequence, the systemic reform initiatives have emphasized agreement on goals for student learning based on rigorous content and performance standards. The theory in the late 1980s was that reform should be coherently organized around content standards—hence the important role of documents enunciating nationally agreed-upon standards in each "basic" school subject (Raizen 1996).

Nationally developed standards were preceded and followed by state curriculum guidelines, often based (at least superficially) on the NCTM *Standards* documents and the Project 2061 *Benchmarks*. (One state, for example, adopted the *Benchmarks* in their totality, illustrated some of them with curriculum and teaching examples, and invited teachers to flesh out the rest in a similar manner.) As of December 1994, 32 states had mathematics frameworks, 32 had science frameworks, and 10 had combined science-mathematics frameworks standards. These state guidelines vary greatly in the level of detail they provide; their length ranges from 20 to 500 pages (Blank and Pechman 1995).

However, as Ravitch (1995, p. 136) points out, "what seemed logically obvious was not necessarily politically feasible." Despite the seemingly whole-hearted acceptance of the NCTM *Standards*, opposition has formed in some quarters to standards-led reforms, sometimes making strange bedfellows of individuals and groups opposing national or state standards. Some were concerned with what they saw as the imposition of a particular set of attitudes and values; others evoked the unfairness of performance standards unaccompanied by program and opportunity-to-learn standards needed by poor and minority students; still others questioned the assessment of the extent to which standards would be achieved through a national test or nationally comparable state tests. Thus, while implementation of the NCTM *Standards* and, presumably, of the American Association for the Advancement of Science (AAAS) *Benchmarks* and newly released *National Science Education Standards* (NRC 1996) continues to move ahead through federal and state initiatives, the long-range future of the standards movement in American education is a matter for speculation.

Partnerships and Collaboration. A third lesson from the 1960s—still played out today in the standards debate—revolves around the power and authority to generate and institute reform. Who is involved in a given innovation, and what role does each party play? The earlier reforms involved scientists and teacher leaders, largely bypassing the education establishment—that is, school, local district, and state administrators. As a result, these reforms were perceived as having been imposed from "outside" the system—by scientists or mathematicians interested in their fields and by the federal government interested in pipeline issues (that is, in the adequate size and quality of the technical manpower pool vis-à-vis the technological challenges posed by the Soviet Union.)

This "elitist" approach appeared to work only in some contexts, more often where curricula and teachers were already reasonably strong, thus leading to the phenomenon of "the rich getting richer." Lack of widespread success could fairly be ascribed to the insufficient involvement (other than of outstanding teachers) of the education system itself—particularly of administrators at the state, local, and school levels—and also of those who had a stake in education, including parents as well as the private sector—that is, the businesses and industries that were the ultimate recipients of a school's graduates. Failure to involve the public and the resulting lack of public understanding was held, at least in part, responsible for such failures of 1960s reforms as the "New Math." However, a careful analysis of that example also shows these influences: opposition from within the mathematical community, inadequate training of elementary school teachers, shallow interpretation by commercial publishers, and a general shift of educational priorities away from mathematics and science education less than a decade after the launching of Sputnik.

In the meantime, the role of some actors increased, particularly that of the states, as noted in chapter 2. This led to an emphasis in current reforms on partnerships and collaboration, given life through requirements put forward in federal program announcements and guidelines for proposals—from NSF and the U.S. Departments of Education and Energy, among others.

Dissemination. The fourth lesson from the 1960s concerns the process of dissemination and implementation of educational reform initiatives. We have already discussed the shift in perspective from a linear implementation model to models that involve diffusion through a social system facilitated by linkers and where the ultimate users are conceived of as problemsolvers. In addition, however, greater attention is being given to means of dissemination. This new attention is in response to what was perceived to be the insufficient impact of the 1960s curriculum development projects, and in recognition of the need to inform and involve all sectors interested in education.

Four of our cases exemplify this in different ways: the Kids Network as one of the original Triad projects requiring collaboration and contribution from a distributor at the outset; the American Chemical Society's continued investment of royalties from the ChemCom text in teacher training; NCTM's unprecedented public relations campaign associated with the launching of the *Curriculum and Evaluation Standards*; and the carefully crafted *Science for All Americans* volume produced by Project 2061, which proved to be so appealing that Oxford University Press reprinted it after the original edition published by AAAS (1989) sold out.

These four characteristics of current reform efforts—systemic reform, standards, collaboration, and dissemination—are, of course, connected. Systemic initiatives are built around the creation of content standards and call for partnerships to help bring about the desired reforms. Consensus on goals and support for standards depend on widespread dissemination, which in turn is helped by enlisting many partners and collaborators in the effort. As we turn next to look at how some of the current conceptions of reform are reflected in the eight innovations we studied, we note the combination of influences, with one or another predominating but the others often noticeably present as well. The organization of the next sections follows the "lessons learned" categories, but, as warranted, the use—deliberate or inadvertent—of alternative models for bringing about R&D–based educational improvement is also considered. Generally, we focus on one or two of the cases that illustrate a given theme most cogently.

Systemic Reform

Even though all were launched before the notion of systemic reform gained currency, at least four of our eight cases can be considered systemic in some way. In fact, one of them—the California science education reform—is sometimes credited with setting the pattern adopted by later systemic initiatives. Project 2061's approach, while proceeding in more linear fashion, also is built around a vision of science education further detailed in *Benchmarks* and augmented by the six school district centers. Project 2061's Blueprints were further designed to address the alignment of curriculum and various other educational elements to make the vision a classroom reality. NCTM, after producing its *Curriculum and Evaluation Standards*, turned next to a consideration of teaching standards and then to assessment standards, thus dealing with three major reform elements. And while the focus of the Urban Mathematics Collaborative is very much on teacher networking, the sponsors hoped for improvement across a whole district, even if participation was voluntary.

California's Example

Ravitch (1995, p. 132) suggests that California pioneered systemic reform under the superintendency of Bill Honig, who became the state's education superintendent in 1982. Indeed, the California case study documents the reformers' clear intent to address all components of the state's system at once. Although the construction of a new vision for science education through the development of an innovative state curriculum framework led the reform, several parallel efforts were undertaken even as the framework garnered national acclaim. The view was that "You can't fix anything unless you fix everything" (Atkin et al. 1996c, p. 19). Textbook publishers were encouraged to produce compatible materials—with the not-inconsiderable incentive of a textbook approval process by a state with 5 million students. Existing teacher networks were strengthened to serve both as dissemination and implementation vehicles; California's institutions of higher education were encouraged to strengthen their links with the schools. External testing authorities, such as the Educational Testing Service responsible for the widely used National Teacher Exams, were asked to bring their tests in line with the *Framework*. California itself undertook a massive effort to develop a whole new range of assessments in line with the *Framework*, later abrogated by the governor for both fiscal and political reasons. While key elements of the system all were being considered, however, the California reformers took an ecological perspective. As the study notes:

> These manifestations of reform were only the most visible part of the story. Beneath them was a particular philosophy of change—comprehensive rather than piecemeal, facilitative rather than prescriptive, built on identified strengths in the system rather than focused on weakness (Atkin et al. 1996c, p. 17).

> Interviews with California education leaders such as Bill Honig, Tom Sachse, Elizabeth Stage, Kathy DiRanna, and Kathy Comfort, reveal that while systemic alignment of educational improvement efforts was embraced enthusiastically, a centralized approach to systemic reform was never seriously considered. One reason is that a top-down approach to systemic reform in a place the size of California would involve an impossible and intolerable level of micromanagement.

> The alternative vision of systemic implementation that emerged was less a totally grassroots approach than a blending of centralized and decentralized initiatives. It can be characterized as consisting of three elements: (1) building a consensus about a general set of priorities for improving California science education among key constituencies, (2) designing an implementation structure flexible enough to be responsive to the evolving needs of the reform, and (3) employing that structure to put the resources and incentives in place that would aid educators at all levels in constructing their own path to the shared goals (Atkin et al. 1996c, p. 20).

At the height of their initiative, the leaders of California's systemic reform seemed to have successfully negotiated one of the oft-discussed dichotomies of educational reform: whether it should proceed top-down or bottom-up, from the center out or through collaboration and networking.

Project 2061

Project 2061 also conceived of educational change as having to address, at least at some level, all the factors that were likely to affect its implementation. Foremost were the vision and goals set forth in the key documents the project had produced, *Science for All Americans* and *Benchmarks*. In an organizational attempt to match appropriate experience to dealing with the several educational components, there was a division of labor among the central project staff in Washington, teachers in the six school district research centers that had evolved from the model sites, center directors, and outside consultants, with quite different tasks assigned to each group.

The leaders in Washington had developed the vision and were in charge of keeping it safe from dilution and distortion; teachers at the model sites, and later those implementing *Benchmarks*, were to move the curriculum forward in the desired direction. To ensure reform of the whole system based on the project's vision, blueprints for implementation were commissioned from experts to deal with teacher education, materials and resources needed for the new curricula, assessment of student learning, curriculum linkages to other school subjects, school organization, family and community involvement, and the role of business and industry. Under current development is the series *Designs for Literacy* to help educators implement Project 2061. The series is to include a discussion of goal-directed reform, a description of the design approach in curriculum reform, advice on undertaking radical curriculum reform, a systematic process for configuring K-12 curricula, and references to relevant research.

Such divisions of labor seem eminently sensible, with each group doing what, by virtue of its knowledge and experience, it seems best equipped to do. In actuality, however, it did not always prove to be so sensible. As the case researchers report:

> Because the worlds of 2061 are physically separate, communication among participants has been crucial for smooth and efficient progress . . . When the flow of information between Project 2061 staff, consultants, center directors, and center participants has been sporadic, misunderstandings have arisen . . . Lack of regular communication among project participants across spheres makes it all the more difficult to appreciate each other's role in the reform effort (Atkin et al. 1996b, pp. 222-23).

One can expect similar tensions in any reform effort broad enough in scope and participants to call itself systemic; one might hope that any such effort has a long enough time horizon, as does Project 2061, to resolve such tensions successfully in the common interest of improving what goes on in the science or mathematics classroom. Indeed, Atkin et al. (1996b, pp. 223-24) go on to note:

> As with most aspects of 2061, this lack of articulation across worlds [of the Washington staff and of the schools] and across participants' roles in the reform has changed over time. Over the years, it appears that participants have learned to accept the existence of these different worlds, have become more sophisticated in their negotiations among them, and have developed a greater appreciation for the work of others. A center director . . . described how his understanding evolved: "We expect change now. Previously we resented it. We take things in stride, expect a fluidity in vision. We're going to do what we're going to do in Georgia, but it's within the confines of the national vision . . . We don't just take things from the top. There's dialogue. We want to further the national goals and be a reality check."

The Power of Metaphors

California may have provided the model of systemic reform subsequently embraced by the federal government (a claim made by Ravitch 1995, p. 146), but the language associated with the systemic reform movement resembles much more that of Project 2061, a language that has taken on the hard edge of engineering—design, blueprint, benchmark—rather than language appropriate to ecological systems.

In fact, the metaphor of a precisely engineered system rather than one that evolves slowly in response to its ecological makeup of loosely coupled subsystems (Weick 1979) pervades current reform rhetoric. For example, in its program solicitation for the first statewide systemic initiative, NSF (1990a) listed 10 components of systemic change: organization, resources, preparation of teachers, continuing professional development of teachers, curriculum content and learning goals, delivery of instruction, assessment of student achievement, facilities, articulation with the system, and accountability. By 1995, these had become the "drivers" of reform (NSF 1995).

Especially prominent was the goals orientation, as exemplified by federal support for the development of standards documents at both the national and state levels. Indeed, curriculum standards were to drive systemic reform—that is, tests and textbooks were to be aligned with the standards, as was teacher pre-service and in-service education. The assumption on which 16 states were funded to create science and mathematics frameworks based on the national standards is the following: Frameworks are created, disseminated, implemented, and then have their effects in changing student performance. The fragmented (not to say chaotic) policy system that had traditionally governed education and education reform was to be replaced by a coordinated set of policies organized around a set of clear student outcomes (Fuhrman and Massell 1992). The outcomes were to be defined through performance standards for students and the schools themselves monitored through accountability systems—all terms borrowed from quality controls applied in systems engineered to provide tightly specified goods or services. This causally linked, mechanistic model of designed change is highly reminiscent of the RDD&E model, but takes its language from industry rather than the research community.

Eisner (1992) presents a convincing case for the power of underlying curriculum ideology to affect what happens in schools. He remarks:

> When the language of industrial competition is used to make a case for particular educational aims, "losing our competitive edge," our conception of the mission of schools is gradually shaped in industrial terms. The school becomes viewed as an organization that turns out a product "a student" whose features are subject to the same quality control criteria that are applied to other industrial products (p. 303).

This approach is not new to American education. In his seminal study, Callahan (1962) analyzed the introduction of Taylorism into education around the turn of the century. Schools were perceived to be inefficient, while at the same time they were expected to do more. If the outcomes of schooling could be defined in terms of standardized goals, then schools could be run more like productive industrial settings, and teachers' performance managed and judged accordingly. In part, Dewey's formulation of progressive education—which saw the purpose of education to be the cultivation of unique talents and teaching as part artistry and part curriculum development in context—was in antithesis to "the cult of efficiency" (Callahan's phrase). As Eisner (1992, p. 313) notes: "It requires no great insight to recognize that these polarities concerning the method and purpose of education are still salient today."

Standards as Goals for Reform

Many difficulties attend the ambitious agenda posed by systemic reform: the complexity of the American educational system with its tradition of local control; the web of mandates—sometimes conflicting, seldom coherent—that emanate from the different governance levels impinging on any given school; the vagaries of funding for the public schools and particularly of any reforms; the changes in educational priorities that hardly allow an innovation to be implemented before a new one is on deck; the large number of stakeholders; the demographic changes in the makeup of the school population; and more (Cohen 1995). Perhaps one of the most critical issues in advancing systemic reform is to get agreement on general goals for science and mathematics education: what students should learn and be able to do. This section discusses the development of standards documents as the vehicle for developing the sought-for consensus.

Goal-setting documents are not really a new development, except at the national level. A number of states have had curriculum guides in place for many years; these may be quite detailed and prescriptive as has been the case in New York since the establishment of the state's educational system in the 18th century. As another example, California frameworks have been published regularly since the early 1960s, but both their rhetoric and their use have changed considerably during the current reform era. As early as 1985, California used its then newly developed mathematics framework to judge the adequacy of the elementary mathematics textbooks submitted for state adoption (Webb 1993). In fact, this framework is considered by some the forerunner of the landmark NCTM *Curriculum and Evaluation Standards* published in 1989. To quote a writer of the NCTM *Standards* (McLeod et al. 1996, p. 39): "Certainly the 1985 *California Framework* was one of the documents that was used in helping to formulate the NCTM *Standards*. It was something that everybody in all of the groups was familiar with and looked at for help in thinking about what the *Standards* might contain." In some cases, even the terminology was the same. For example, "mathematical power" was a phrase from the *California Framework* that was also emphasized in the Standards. As John Dossey [a leading mathematics educator] noted: " . . . The *Standards* get credit for it, but it was in that *California Framework.*"

Why Nationally Developed Standards?

While goal-setting documents have a long history at the state level, at the national level, statements of national goals represent a major shift in educational policy. A *Nation at Risk* (National Commission on Excellence in Education 1983) played a significant role in this change through its call for more national attention to educational improvement. The inflammatory rhetoric of this report, accompanied as it was by a well-publicized campaign of regional meetings, got the attention of the media, the business community, and the public—an "unprecedented" response, according to Ravitch (1995). All of the attention helped in resisting the Reagan administration's initial plans to reduce the federal role in education: plans which included the abolition of NSF's precollege programs and of the U.S. Department of Education (a recommendation revived by some members of Congress 15 years later).

One of the first reactions on the part of states was to increase requirements for high school graduation, as the reports urged. Between 1982 and 1992, student enrollment in algebra 1 increased from 65 to 89 percent, in algebra 2 from 35 to 62 percent, in calculus from 5 to 11 percent, in biology from 75 to 90 percent, in chemistry from 31 to 55 percent, and in physics from 14 to 23 percent (Blank and Gruebel 1993). Moreover, Porter et al. (1993) found that the increased student enrollment did not make for "watered-down" courses, as some had feared; the college-preparatory mathematics and science courses essentially maintained their integrity. Nor did the increased enrollment in such courses lead to greater dropout rates, another concern voiced at the time.

Another finding, however, was that these courses remained quite traditional, unaffected by various reform initiatives. When increased enrollment in mathematics and science courses seemed an insufficient means for elevating students' achievement in these fields, the notion of systemic reform to address various components of the education system, not just one, began to take hold. The recognition that systemic reform of education would be assisted by national statements of educational goals, or standards, helped to overcome, at least for a time, the long-standing fear of federal control of education through national curriculum policy.

It may have helped that the very first set of standards to come along were not federally sponsored nor considered elitist. Rather than being led by mathematicians and scientists, as in the 1960s, leadership in the effort to improve mathematics education in the 1980s was taken over by NCTM, a professional association of mathematics teachers and teacher educators. As Atkin (1994) notes, NCTM was a prime factor in the current effort to base reform on well-articulated standards. Part of the reason for NCTM's success was fortunate timing; as Ravitch (1995, p. 57) notes, "At the very time that governors and other political leaders wondered about the feasibility of voluntary national standards, there were the NCTM *Standards* as an example for emulation."

What Are Standards?

As we have noted, "standards" is a term with multiple meanings. One meaning suggests an accountability emphasis, or the use of standards as criteria to assess quality; as Atkin (1994, p. 82) puts it, "'Standards' has the kind of bite politicians like." The accountability emphasis was very much in the spirit of the 1980s, and this interpretation certainly had a big impact on the ready acceptance of the notion of standards among federal and state policymakers, including the nation's governors who had articulated the national education goals.

Among professionals in the fields of mathematics and science education, however, the term has come to reflect a different meaning—like a banner raised in battle, standards are a rallying point embodying the vision and ideal to be promoted and pursued. The NCTM case study describes how the emphasis shifted in the mathematics education arena from the accountability notion to a focus on standards as a vision of the future. The meaning of the term has become even more clouded as various constituencies have expanded the concept beyond content standards to performance standards (to what degree of proficiency are students to learn the given content) and opportunity-to-learn or program standards (what do schools need to provide to enable students to learn the content at the requisite proficiency level).

The multiple meanings of the term "standards" have been a blessing and a curse. Many authors (see, for example, Atkin 1994) note how the multiple meanings have enabled "standards" documents to attract support from disparate groups. In the case of mathematics, Apple (1992) discusses the NCTM (1989) *Standards* as an example of a "slogan system" in education—vague enough to allow many supporters, specific enough to offer something to practitioners, and engaging in a special way that charms the reader into believing in the vision of the future that the document presents. The dangers of these multiple meanings have also caused many authors to worry about how the "standards" documents might be used in ways that are not in the best interests of mathematics and science education. As Kilpatrick and Stanic (1995, p. 11) point out, "There is a price to be paid when slogans are stretched to cover multiple agendas."

Roles of Standards Documents

We have briefly described some of the impetus behind the movement toward national standards: the development of systemic approaches, the perceived needs of standards to guide reform, the multiple meanings of the term standards, and the shift from state to federal leadership in reform documents. In this section, we deal with how the nature of reform documents themselves has changed.

As noted by Raizen (1991), in the 1960s era, fresh from their successes during World War II with mega-projects in science and technology, leading mathematicians and scientists provided direction for educational reform through major curriculum improvement projects. The shift in educational priorities a decade later, the perceived lack of success of the 1960s reforms, as judged by 1970s standards, and the consequent withdrawal of many leaders in mathematics and science from efforts to improve education, led to a return to traditional patterns of instruction. Some even go so far as to claim that the failure of "New Math" was a factor "leading to a 'back to basics' movement and the spread of minimum competency testing in dozens of states" (Ravitch 1995, p. 49).

At the same time, behaviorist psychology began to influence educational philosophy as well as educational and other public management.[2] Curriculum was stated in fine-grained behavioral objectives. The emphasis on testing for basic skills further reinforced the specification of curriculum guidelines in terms of measurable objectives. States and districts, as well as curriculum developers and textbook authors, were expected to specify these objectives in considerable detail.

NCTM. The style of the 1989 NCTM *Curriculum Standards* document was a radical departure from the practice of specifying behavioral objectives in great detail. This was well-captured by a state supervisor of mathematics (McLeod et al. 1996, p. 116):

> Most people think of what they're supposed to know and be able to do in terms of a list: "Let's itemize the list of things we're supposed to know and be able to do." One of the brilliant characteristics of the *Curriculum Standards* is that the grain size is big enough that the bullets aren't damaging; they cannot be treated like behavioral objectives. Brilliant move! They're not sweepingly general; they can't be rejected because they're too general. But they're not at the small grain size where you'd have 150 per grade span, [the grain size] that people are used to.

There are, of course, additional reasons why the NCTM *Standards* have become so widely accepted. Although the standards effort was the largest project ever undertaken by the organization, an earlier policy document, NCTM's (1980) *Agenda for Action*, provided an opportunity to see how a document could provide a useful focus for NCTM's professional activities. All during the early 1980s the focus of NCTM's national and regional meetings, as well as of its professional development and publication committees, was on implementing the points in the *Agenda*. Although NCTM's investment in the *Agenda* was relatively small, the dividend in terms of attention from federal agencies, teacher education organizations, and publishers was relatively large. When NCTM developed the *Standards* during 1987-89, there was a base of experience with the *Agenda* on which to build.

Moreover, as Crosswhite (1990) points out, the NCTM *Standards* were more evolutionary than revolutionary. The writing teams that produced the *Standards*

---

[2] The late 1960s were the heyday of management-by-objectives in the federal government, introduced by Secretary of Defense MacNamara.

were structured to include the different constituencies represented in the organization: teachers; mathematics supervisors; teacher education faculty; and others with expertise in research, technology, or other areas (Crosswhite et al. 1989). The NCTM leadership was careful to include older, experienced members on each writing team so that the final document would convey a vision of the future that was tempered by realistic knowledge of past reform efforts. The writers' sensitivity to the concerns of teachers and others with more traditional views was an important factor in the document's wide acceptance in the field.

NCTM also tried to broaden the appeal of the *Standards* by reaching out to other organizations in the mathematical sciences. Through the establishment of the Mathematical Sciences Education Board, NCTM was able to gain support from all the major professional associations in mathematics (Crosswhite et al. 1989). As Massell (1994) points out, mathematics may be unique among all school subjects in terms of its relative unity all the way up the educational ladder from kindergarten through graduate school. Certainly the mathematical sciences are much more unified than the natural or social sciences. The difficulty that other groups have had in developing standards (Donmoyer 1995, Myers 1994) only serves to amplify this distinctive characteristic of the discipline of mathematics.

California. A similar approach, with respect to both process and style, was taken by the *California Science Framework*. Honig, at the time the chief state school officer, said: "The *Framework* was to give focus [to the California science education reform] without giving explicit direction" (Atkin et al. 1996c, p. 21).

Many teachers may have expected more precision with respect to curriculum content in the new *Framework*. They didn't get it. To bridge the gap at the elementary level, staff developers and teacher leaders introduced the notion of each school developing a content matrix based on the *Framework* to create an articulated, coherent K-6 science program. This was very much in the spirit of teacher/implementors as problem solvers, but it exacerbated teachers' workloads, already considered burdensome by many of them. By specifying curriculum goals in a larger grain size, the reformers were following their own constructivist views of learning (see chapter 3), which they applied not only to K-12 students, but to the teachers of these students as well. They were not trying to specify a set of behavioral objectives for teachers to address; they recognized that teachers, like other learners, would take the ideas presented in the reform documents and reconstruct them in their own ways. Although such an approach raises concerns in terms of the fidelity of the reform effort, the approach is consistent with the view of the learner that is often espoused in reform efforts in science and mathematics education, not only in the United States but in other countries as well. Such an approach is also in line with the view of teachers as professionals who should have the power to define both the content and the instructional approaches for the subjects they teach, as discussed in chapters 3 and 4.

The risks in such an approach have been outlined by Donmoyer (1995), who laments the fact that the "big tent" approach of standards documents—he was commenting specifically on the November 1994 draft of *National Science Education Standards* produced by the National Research Council—leaves all the hard decisions for later in favor of political buy-in through vagueness. Indeed, his fears may be well-founded: Even where the *California Framework* seemed to have been implemented, the resulting program sometimes proved insubstantial, in part because of the myriad possibilities of interpretation. This phenomenon is further documented in two sets of studies (Marsh and Odden 1991, *Educational Evaluation and Policy Analysis* 1990) of the implementation of the *California Mathematics Framework* (California Department of Education 1985) which seemingly come to opposite conclusions. Knapp (1995) interprets the contradictory findings as follows:

> . . . the curricular policy was successful during its early years in activating a large number of actors in regional networks, districts, and schools on behalf of an ambitious mathematics reform. Various factors . . . converged to encourage large numbers of teachers to pay attention to the policy and realize it in the classroom. There the policy has met with more mixed success—so far. Teachers' efforts to put the policy into practice have succeeded in changing the more easily altered aspects of practice but have been limited by the teachers' grasp of and beliefs about mathematical knowledge itself, their capacity to visualize how a new conception of knowledge translates onto teaching, and by the momentum of teaching traditions.

The need to provide adequate guidance for teachers in interpreting standards has led the American Federation of Teachers, a strong advocate of rigorous standards, to promulgate as one of its criteria for effective state standards: "Standards must be specific enough to assure the development of a common core curriculum . . . " (AFT 1995), which the organization sees as entailing specification by grade levels or age bands, clear and specific enough to guide curriculum at the local level, leading to all students learning the defined core curriculum, and alignment of textbooks and tests.

Project 2061. Project 2061 sought to avoid the dilemma of inappropriate or ineffective interpretation at the classroom level. It addressed the problem through "backmapping" in order to translate *Science for All Americans* into curricula that could be implemented by teachers. Backmapping consisted of analyzing the ultimate learning outcomes described in this document to identify the sequences and approximate grade levels in which the ideas should be taught. This turned out to be very arduous work, and not all project staff or team members at the six centers agreed with Roseman, the project's curriculum director, that the backmapping process was essential. The case study comments (Atkin et al. 1996b, p. 166):

After a painstaking process, the teams eventually produced maps for many of the learning outcomes. In a proposal to NSF, the project reported that while sites worked independently on backmapping tasks, analysis of their work showed remarkable agreement on outcomes, or "benchmarks" as they have come to be called. These maps became the basis for the benchmarks (though the two are not identical) which, notably, were initially conceived as an internal product.

With the favorable reception of the NCTM *Standards* and the drive to create standards in all major school subjects, this internal product became the nationally published and disseminated *Benchmarks For Science Literacy* (AAAS 1993), providing some 855 knowledge goals to serve as a more detailed key in designing curricula faithful to the vision of *Science For All Americans*.

Kids Network and Voyage of the Mimi. A yet more detailed approach is illustrated by both the Kids Network and Mimi cases, where the attempt to maintain the fidelity of the innovation led developers to be quite specific about what teachers should do. Both provide special guides for teachers to use in implementing the materials. In the case of Kids Network, these contain background information, weekly lesson plans for core activities and extensions, and a software manual. Our case study gives the example of one unit (deemed quite typical) that contains a 48-page teacher guide: The guide starts with a 14-page introduction to the program and overview of the unit; this is followed by the six-week lesson plans, each of which is divided into one or more class sessions including objectives, preparation and materials, activities, homework, and teaching tips and background.

The Mimi overview guides, provided in looseleaf notebook format to allow teachers to make their own additions, are somewhat less specific. They contain a mix of information and activities for each episode and expedition, as well as some preview and follow-up questions. The guide for the second series also contains factual background material, activities for the students, and references and a resource list—over 200 pages in total. Yet, despite detailed assistance for the teacher in an attempt to preserve the original spirit and philosophy of these innovative curricula, or perhaps in part because of it, teachers followed their traditional approaches. Mimi's case study researchers write:

> Mimi did not substantially change the way teachers taught science—science instruction continued to be short, mostly talk, and teacher-directed. For example, to move the discussion along, preferred answers were short and specific, leaving little time for reflection, the airing of different perspectives or expansion of the topic. Additionally, hands-on activities—although more numerous than previously—followed scripted material (Middlebrooks et al. 1996, p. 505).

The Kids Network case researchers observed similar discrepancies among their vision, the publisher's desiderata for materials that would be widely adopted, and the classroom practices of teachers. Their explanation, more amply discussed in chapter 7, is one of "time warp": that is, the differences in pedagogies espoused by researchers, reformer/developers, publishers, and teachers come about through adherence to newer or older teaching paradigms, with replacement of the old by the new taking place at different rates for these different groups.

In sum, when the reform documents are written in a larger grain size, teachers and others will have varying interpretations of the meaning of the reform, often seeing superficial change in practice as implementation of fundamental reform (Cohen 1990). When teachers, in order to preserve a reform's goals and consistency of vision, are given the kinds of detailed specifications they are used to, they may well continue their traditional practices. In either case, teachers may appear conservative because they know the effects of what they have been doing, perhaps for many years, and they are not convinced of the efficacy of the new; or they may truly welcome new practices but not recognize the changes that need to be made or how to make them. There also may be many pressures on them to continue past practice—expectations of administrators, parents, and other influentials in the community; external testing programs; university admission criteria. Such considerations have given rise to the view that reform needs to engage not just scientists and teachers, as in the 1960s, but all the actors whose views impinge on the classroom.

Partnerships and Collaboration

Partnerships and collaboration have been considered intrinsic to systemic reform. More commonly, however, partnerships to foster improved mathematics or science education are simple alliances between a school and a local business, museum, hospital, university, or other "science-rich" institution. By 1991, the number of school-business partnerships or alliances devoted to mathematics and science had grown to 500, up from 60 some five years earlier, according to the Triangle Coalition for Science and Technology Education (1991). The Coalition defined a typical alliance as including partners from (1) a school system; (2) business, labor, industry, or chamber of commerce; and (3) science or engineering departments of universities, science-oriented professional societies or government agencies, or a local museum.

While early partnerships often did not consider themselves to be playing a central role in reforming classroom instruction (Blair et al. 1990), this began to change as the notion of partnership and collaboration became part of the systemic reform agenda. For example, in 1990, NSF solicited proposals involving private partnerships to improve science and mathematics education. mainly at the precollege level. The primary intent was to blend "the scientific, technical, organizational, and management skills of the non-educational world with the knowledge and expertise of educators" (NSF 1990b, p. 1). Significant participation by classroom teachers was of particular interest to NSF; possible objectives for proposed projects included improving the teaching of mathematics and science, implementing integrated science courses and emphases on science applications and laboratory components, and motivating students in science and mathematics (and possibly teaching careers), as well as improving their preparation.

Private foundations, too, found the notion of partnerships and collaborations promising as a reform strategy. By 1984, the Carnegie Foundation had made 20 grants to link "science-rich" institutions to schools. Atkin and Atkin (1989) gave a positive report to the Foundation on these alliances but also warned about their fragility and the sometimes unfocused nature of their activities. The Carnegie Foundation also supported the Triangle Coalition for Science and Technology Education to foster (beyond its own funding) partnerships between the private sector, particularly business and industry, and schools.

But perhaps the most ambitious project in collaboration was funded by the Ford Foundation, starting in 1985, the Urban Mathematics Collaborative project, one of our eight cases. Although the focus was on empowering teachers by increasing interaction among them, the collaboratives also were to encourage interaction between teachers and mathematics professionals in industry and universities and between teachers and mathematics education reformers. The UMC project is unique among the eight projects we studied only in its bald assumption that collaboration among teachers and partnerships with business, higher education, and mathematics education leaders came *before* and not after a vision of the needed reforms was created. The UMC case study researchers write:

> The project was developed on the basic principle that well-connected and professionally active teachers would be able to improve upon the poor mathematics performance of inner-city students. The central goal of the UMC project was to empower teachers of mathematics by increasing the interaction among teachers, between teachers and other professional users of mathematics in business and higher education, and between teachers and others involved in mathematics education reform (Webb et al. 1996, p. 248).

This stands in marked contrast to the NCTM, AAAS, and California efforts, all of which—systemic though they are—started with a vision document, to say nothing of the four curriculum-focused materials development projects. Perhaps it was fortuitous for the collaboratives that the NCTM *Standards* made their appearance when they did. As the UMC case study notes (Webb et al 1996, p. 272): "The *Standards* appeared at an opportune time—collaborative directors and coordinators were seeking more focus after a productive phase promoting involvement and awareness." This certainly seems in line with Fullan's (1993, p. 28) dicta that "First . . . one needs a good deal of reflective experience before one can form a plausible vision. Vision emerges from, more than it precedes, action . . . Second, *shared* vision, which is essential for success, must evolve through the dynamic interaction of organizational members and leaders." Quoting again from the UMC case study:

> All of the collaboratives in some way supported activities to make mathematics teachers aware of what changes were needed in the mathematics curriculum and why such changes were necessary. For example, the collaborative coordi-

nators and directors gave presentations about the NCTM *Standards* while the documents were in draft form . . . groups of teachers [in some of the collaboratives] reviewed drafts of the *Standards* and sent their comments and concerns to the writing team. District mathematics supervisors from collaboratives were afforded the opportunity as a group to discuss and think about means of implementing ideas from the *Standards* . . . As a result of the activities of the national UMC network, at least one person at each site had a fairly extensive knowledge of the *Standards* . . . (Webb et al 1996, p. 272).

In the case of California's science education reform, the strategy envisaged widespread involvement of many communities from the outset. It started with the very process of creating the 1990 *Science Framework* and the nature of that document. A very deliberative process was set in motion as the framework was planned and written; the writing committee included representatives of organizations and state leaders who had already begun efforts to improve science education. To help forge consensus, the actual writing of the document had to bring together these many able and influential actors. The content of the *Framework* itself also was to contribute to building a broad base for science education reform, as our case study notes:

> [The 1990 *Science Framework* was intended] to be "provocative," to stimulate serious consideration and extended discussion . . . The relatively open-ended text required interpretation to be useful. It also required involvement not only by teachers, textbook publishers, and school administrators, but also by teacher educators, parents, and school board members. In turn these participants in the statewide effort derived direction and legitimation from the *Framework*. This relationship between a key document and its many audiences is sometimes subtle and easy to miss (Atkin et al. 1996c, p. 18).

Indeed, a great deal of energy was created around the state. But the main vehicle for implementing the *Framework* was the use and expansion of existing teacher networks and the building of new ones, not unlike the mode of the urban mathematics collaboratives, and with some of the same success as well as dangers. According to the case study, about a third of California's 6,000 elementary and middle schools have paid the annual fee of $2,500 to belong to these networks dedicated to improving science education in line with the *Framework*; 300 of 800 schools are involved at the secondary level. Thus, teachers were seen as both the most critical and strongest element in the whole reform effort. In the words of the case study, the reform's leaders believe that:

> Articulating goals for the science program (in California's case, the new *Science Framework*) is not enough. The people who subscribe to the broadly stated goals must understand their meaning and figure out how they want to translate them into curriculum. This takes resources, lots of resources. But the prime one is creating opportunities for teachers to talk to one another (Atkin et al. 1996c, p. 112).

In contrast to Project 2061, California's strategy gave primacy to teachers' consideration of what they ought to teach in the classroom—the teacher as problemsolver regarding science instruction—much as the teachers in the Urban Mathematics Collaborative project were to figure out for themselves how to improve their mathematics instruction. And as in the UMC project, rigorous content sometimes took a backseat. The California case study researchers conclude, in part:

> [A] major issue for California has been the problematic quality of the science that is being taught. In many classrooms—particularly at the elementary school level but also at the secondary level—the actual science teaching is often episodic, superficial, and inconsistent with the *Framework*. Elementary school teachers generally do not have strong science backgrounds. In choosing to build the reform around the most committed of current teachers, there was a tradeoff between the resulting conception of science, even its accuracy, and gaining the emotional commitment of large numbers of teachers. Leaders of the reform in California recognize the seriousness of the problem but believe involvement and commitment by teachers come before what they see as the even more difficult task of improving their science backgrounds (Atkin et al. 1996c, p. 113).

Dissemination Strategies

The NCTM *Standards*

California's approach of involving many communities in the process of creating the state's 1990 *Science Framework* to facilitate its acceptance was amplified in the case of the NCTM *Standards*. The review process for the *Standards*—write draft, circulate for review for a year to membership and outsiders, use draft as a focus for regional and national meetings, etc.—provided an extended opportunity for dissemination and for the membership to feel an investment in the project. This was a change indeed from the 1960s style of curriculum development, where cycles of revision generally involved a small core group of scientists or mathematicians and teachers.

But perhaps an even greater departure was the use of a public relations firm to make the *Standards* widely known. The case study tells the story best (McLeod et al. 1996, pp. 15-16):

> On March 21, 1989, the National Council of Teachers of Mathematics (NCTM) held a press conference in Washington, DC, to release the *Curriculum and Evaluation Standards for School Mathematics* (NCTM 1989). The *Curriculum and Evaluation Standards* had been under development for almost five years . . . and public interest was high . . . As Marilyn Hala, an NCTM staff member, recalled: "This is the first time I ever remember that we had a bank of six or seven video cameras in the back of the room. There were probably 200 people at the press conference in the Willard Ballroom. It was quite impressive! You had to walk by these members of the press—these cameras—to sit down." . . .

As John Dossey [past NCTM president] put it:

> "That's why I was on the *Today Show* [on NBC-TV], why Shirley Hill was on
> CBS or ABC during that period of time. That's why Sally Ride and Bud
> Wonsiewicz were at the press conference ... [The PR firm] took me on a tour of
> New York City one day; I spent the morning talking with the . . . *Wall Street
> Journal*, the afternoon with the . . . *New York Times*, and the next day on the
> telephone with the *Los Angeles Times*, the *Chicago Tribune*, and a few others.
> Those were not things that we were used to doing."

> The *New York Times* and the *Washington Post*, along with many other newspa-
> pers, reported on the release of the *Curriculum and Evaluation Standards*
> (NCTM 1989) in their issues of March 22, 1989. NCTM President Shirley Frye
> and Past-President John Dossey were quoted on the need for reform in mathe-
> matics education; the newspaper reports focused on the need for reasoning over
> rote learning, the usefulness of calculators and computers, and the emphasis on
> applications. Critics of NCTM, like textbook author John Saxon, were also
> quoted on the dangers of technology and the need for drill and practice. A
> "strong national debate" was expected, according to the *New York Times*.

Another major dissemination effort, "Leading Mathematics into the 21st
Century," was developed by the Association of State Supervisors of Mathematics. Each state (and sometimes each congressional district within a state) had
teams of leaders in mathematics education who organized regional conferences.
Project leaders document contacts with 50,000 to 60,000 people and estimate
that their conferences actually reached twice as many teachers and leaders at the
local level.

It is worth noting that Project 2061 also has been very successful in getting
media attention building on the media's interest in standards. For example, when
Benchmarks for Science Literacy was published in 1993, it was hailed by a number of print and broadcast media as "the new national science standards."

Materials Development Projects

ChemCom. ChemCom represents a highly successful example of dissemination among the curriculum development projects. The textbook has
become popular throughout the country: by 1995, a half-million students were
conservatively estimated to have taken the ChemCom course, based on total sales
of 300,000 copies of the text. Moreover, the text has been translated into several
foreign languages and has been distributed, among other places, in Japan, Russia,
and Mexico and several other Latin American countries; negotiations with China
are in progress. A one-semester college course following the ChemCom model
has been developed (*Chemistry in Context*), as have a series of modules for middle school (FACETS—*Foundations and Challenges to Encourage Technology-
Based Science*).

What has made the dissemination of ChemCom so successful? One factor
was the strategy of investing early in master teachers through residential 10-day

workshops over a period of three years (1988-91). These teachers were then available as a corps of individuals able to staff the expanding demand for teacher workshops. A second factor is the continuing support of its original sponsor, the American Chemical Society (ACS), which continues to plow royalties from the textbook sales back into teacher training. For example, in 1994, $226,000 was spent from royalties and other ACS funds for 10 five-day summer workshops; $220,000 was spent in 1995 for another 10 workshops. Five free newsletters per year are distributed, with a circulation of over 20,000. An ACS team and 20 high school teachers trained 140 Russian teachers; teacher training workshops were also held by ACS in Mexico.

None of the other three materials development projects have reached ChemCom's popularity, nor have they been able to match it in dissemination resources. Yet they have not lacked for innovative dissemination mechanisms. For example, "Mimi fests" are held in various regions of the country, often at maritime or natural history museums; these star Peter Marston, the actor playing the captain of the *Mimi* and, in real life, the owner of the boat. These are attended by hundreds of schoolchildren and their teachers. Diffusion of the Kids Network is helped through having a nationally prominent scientific society as distributor of the materials, as well as having received NSF funding as one of the original Triad projects.

The Precalculus Course. The less planful diffusion of the North Carolina School of Science and Mathematics (NCSSM) precalculus materials through various "elite" teacher networks led the case study team researching this innovation to change the focus of their study from examining the course per se and its enactment in various classrooms to an examination of its development by a group of teachers and its spread through the involvement of these teachers in local and national communities of teachers interested in making change in their mathematics teaching. The Woodrow Wilson Summer Institute for Teachers at Princeton University served as a critical venue for disseminating information about the NCSSM approach to teaching a precalculus course through applications; other venues were provided by Ohio State University, Phillips Exeter Academy, and the Council of Presidential Awardees (formed by the yearly winners and comprising outstanding mathematics teachers). The case study report notes (Kilpatrick et al. 1996, p. 214): "Although the national community of mathematics teachers working to change secondary mathematics courses and teaching was not well-defined, it seemed to provide a powerful source of energy and support for reform." The report goes on to draw a contrast between the approach of the NCSSM teachers and the usual appeals made to teachers:

> American education is beset by people with something to sell. Teachers' mailboxes overflow with catalogs, brochures, and other come-ons. Teachers attending conventions find not only the exhibition space occupied by purveyors of instructional materials but also the meeting sessions filled with speakers touting their latest patented nostrum. NCSSM teachers, too, had something to sell —a course, a book, an approach to mathematics and to teaching—but in their work

with other teachers, they acted like colleagues and not merchants. They seemed unconcerned about converting others to a cause or making a sale. They were neither evangelists nor hucksters. Rather, they wanted to exchange ideas (Kilpatrick et al. 1996, pp. 228–29).

Beyond the collegiality, there also was the lack of pressure and the regard teachers had for their colleagues' knowledge and experience:

> The communities of teachers with whom NCSSM teachers worked had not been formed under coercion. The communities, local to national, comprised self-selected professionals who were seeking change. These teachers were less impressed by advanced degrees than by firsthand knowledge and experience. They returned to the conferences and workshops because they found them valuable and liked the atmosphere . . .

> NCSSM teachers earned their colleagues' respect because of what they had to offer and how they offered it, not because of their names. Although several members of the NCSSM mathematics and computer science faculty, as well as others in the community centered at NCSSM, had become nationally known figures in mathematics education, the atmosphere at conferences and workshops was marked by a high degree of social equality (Kilpatrick et al. 1996, p. 229).

These findings read very much like the literature on diffusion of agricultural practices, which has proved most successful when the social distance between the demonstrator of an improvement in practice (the agricultural extension agent) and the potential adopter (the farmer or rancher) was minimal.

Developer-Publisher Relationships

Most educational innovators—the NCSSM teachers perhaps being an exception—are keenly interested in spreading their reforms to many sites. But no matter how successful and acclaimed their innovations may be, unless there are effective dissemination channels available and interested, access to and acceptance by the ultimate users in the schools will be limited.

After innovative curricula have been successfully field tested, their creators generally seek distributors with capacity to reach teachers, administrators, curriculum specialists, and others influential in textbook adoptions across the country. One problem is that curriculum materials are no longer limited to print material (textbooks, teacher guides, supplementary information) but—especially in science—include kits for hands-on activities as well as multimedia products (e.g., videos, computer software, videodisks, and CD-ROMs). Finding a publisher or distributor willing to take on a new project or product is no trivial task, especially if it includes hard-to-handle materials. Under the best of circumstances, a new product requires a publisher's investment in art, editing, printing, packaging, and marketing. When the projects are innovative and, moreover, consist largely of nontraditional materials, they are financially risky in the eyes of publishers. And they may require a publisher to reach new markets, always a costly venture.

Once a project finds an interested publisher, tensions often develop around two issues: editorial control, and marketing plans and practices. Understandably, developers want editorial content control. They have content expertise and field test data upon which to make decisions. Publishers, for their part, believe they know what the market will buy—i.e., what needs to be done to a product to make it salable—hence they too want editorial control. This tension has played out differently in the four curriculum development projects included in our case studies.

The wisdom of establishing ground rules at the start of a collaboration for how editorial control will be negotiated or decided can be seen by contrasting the experience of projects that kept editorial control of content with those that did not. Consider, for example, the experience of ACS when it was ready to publish ChemCom, its new high school chemistry curriculum. Eleven publishers out of more than the 20 that were invited came to a meeting held by ACS to select a publisher for its ChemCom textbook, especially planned to attract a wider range of students to chemistry. ACS declared editorial control of content as a nonnegotiable condition. Participation by ChemCom staff in planning some aspects of marketing was another condition set by ACS. At the end of the briefing session only one publisher, Kendall/Hunt, which markets primarily via catalogs, remained interested. KH agreed to the conditions. ACS retained content editorial control, and the two organizations cooperated in the marketing arena. In addition, KH agreed to support a series of short awareness sessions which ACS staffed with knowledgeable teachers. These arrangements between ACS and KH have continued through two editions. Work on the third edition is in progress. ACS uses its royalties to develop and distribute supportive supplemental materials such as a newsletter and a final examination for the course, both of which it publishes.

Editorial control and latitude to experiment, i.e., to be responsive to what they were learning from experience, also was important to the NCSSM mathematics teachers who developed the materials that eventually became *Contemporary Precalculus Through Applications* (Barrett et al. 1992). The group had neither the resources nor experience to do its own publishing. One of its members approached an official from the innovative textbook division of Addison-Wesley, a large publishing house. Preliminary discussions did not seem fruitful for either the publisher or the project team. The team felt that a smaller publisher might be more likely to give them editorial latitude. At a meeting of the Mathematical Association of America in 1990, some NCSSM team members discussed their need for a publisher with Barbara Janson of Janson Publications. Janson had never done a textbook. She was willing to give the teachers editorial control with the condition that the book be reduced in length, which the teachers did. The NCSSM project provides an example of how both parties benefit if they can work together with some reasonable mutual understanding and adjustments.

As NCSSM teachers conducted in-service programs related to their project, the need for additional supportive materials became apparent. They prepared a manual on graphing calculators and an instructor's guide, which they thought needed to be part of the published package. The publisher agreed, since the additions made good marketing sense.

The two other curriculum projects, each with technology components, had far more difficult and less mutually satisfactory relations with their publishers. One of them parted company with its initial publisher; the other stayed with its publisher, but not without some frustrating moments for both parties.

The Voyage of the Mimi, followed by *The Second Voyage of the Mimi,* combines television, computer software (videodisk) and print materials to present an integrated set of concepts in mathematics, science, social science, and language arts. The material associated with each Mimi trip is meant to supplement existing curricula in the upper elementary grades. Holt, Rinehart and Winston, an established textbook publisher with a sales force in the field, took on the project and produced the material for the first Voyage in 1984. The company did not have experience packaging and marketing computer software. It had little financial incentive to do so, since the required hardware was not yet widely available in elementary schools. Moreover, Mimi presented a special problem for the sales force. Since the project was clearly cross-disciplinary in orientation, to whom should it be marketed? Was it science? social studies? From what part of the school budget should the money come? From the perspective of the publisher, the marketing challenges and risks posed by Mimi were cause for concern.

Neither side fully appreciated the difficulties of the other. On the one hand, Holt, Rinehart and Winston, with its sales people in the field, felt it was tuned to what teachers wanted—i.e., to what would sell. The company had learned from long experience to avoid marketing products that required anything but very small incremental changes in current practice. On the other hand, innovative projects like Mimi often have new goals that require changes in curriculum as well as teaching practices. In any event, after visiting field test sites, Mimi evaluators argued that the main challenge facing developers was to make their materials as comfortable to as many teachers as possible. Samuel Gibbon, director of the Mimi project, remarked:

> We were comfortable if boundaries around science and scientists got blurry . . . We really wanted kids to imagine a kind of scientific activity that included scientific observation, messing around in the data, looking for patterns . . . We intended to suggest that you could be curious about anything . . . [that] what distinguished scientific curiosity was suspension of belief, questioning of data, challenging of authoritative statements, continuing to keep an open mind about things (Middlebrooks et al. 1996, p. 410).

> In fact, classroom materials—the print materials—are quite conventional . . . [We were] dealing with a publisher who wanted to get things into schools and teachers who had certain ways of proceeding, and [we were] already bold in asking teachers to show videos and use software (Middlebrooks et al. 1996, p. 413).

Gibbon found another publisher for the second Mimi series. This time, instead of a textbook publisher, he turned to a small software company that had never done a textbook project. Wings for Learning, a subsidiary of the Sunburst Corporation, markets educational multimedia packages, primarily software-based, by catalog. For the *Second Voyage of the Mimi*, Sunburst published videotapes and videodisks, student books and teacher guides, software, posters, a newsletter called the *Mimi Experience*, and an array of extension or supplemental materials. The newsletter contains stories and ideas from Mimi teachers as well as additional challenges for students such as Mimi-related contests sponsored by the company. In short, Sunburst had the necessary flexibility and enthusiasm that the larger, traditional publisher seemed to lack.

We noted earlier that the Kids Network was one of the Triad projects supported by NSF. These projects represented NSF's attempt to have major publishing houses participate earlier in the life of the curriculum development projects it was funding; the grant competition required proposals to have a publisher committed to the project ahead of time. The assumption was that providing a fiscal incentive for leery publishers would encourage them to take on new ventures in elementary school science and that having developers and publishers work together from the outset would improve the format and marketability of the final products. However, since the publisher had to agree to match the NSF investment in some combination of money and in-kind services, scientists and curriculum developers with innovative ideas had to scramble to find and get commitments from publishing houses that on the whole saw more risk than benefit from participation. In negotiations with reluctant publishers, project directors were in a weak position.

In the case of the Kids Network project, published by the National Geographic Society (NGS), failure to keep editorial control of content in the hands of the principal investigator, Robert Tinker at Technical Education Research Centers (TERC), and lack of experience in a long-term collaborative relationship with authors on the part of the publisher, affected the project's products. As Dot Perecca from NGS said in an interview with James Karlan:

> It was the first time in the history of National Geographic that we had ever had a government grant . . . Up until that time, we did everything ourselves. We didn't go outside looking for individuals to help support what we were doing. So this was a whole new concept for us, for TERC, for NSF. It was a bumpy road.

The kind of collaboration envisioned by NSF simply did not happen. For one thing, the publisher, visiting classrooms with an eye as to how to make the Kids Network units salable, found that its greatest strength was in its contribution to language arts—yet TERC is a science-focused organization, as is NSF, the funding agency. Nevertheless, NGS saw the computer communication feature as a powerful stimulant for language development. After this and other attempts at collaboration faltered, the relationship was renegotiated. NGS reverted to a standard publishing practice: namely, TERC presents to NGS completed, field tested units, and NGS decides to accept or reject them as it sees fit.

Asked what advice she would give to others considering early publisher-developer collaboration, Perecca said there were not sufficient reality checks by the developer. There is a gap between " . . . pie in the sky ideas for developing something and what the reality was of the need to sell it . . . and that was probably where we had our biggest differences." She gave as an example TERC's decision to put into the kit an expensive piece of equipment to test pH, when pH paper would have done the job. "The kit was broken the first time somebody used it . . . So, there was a lack of reality check of what is salable to the majority."

Developers may not understand what is required of the publisher in marketing a product because they have never had to do it. As Perecca remarked: "They never had to box the product or guarantee it. They never had to replace a product that didn't function. That was a serious issue—you can create the greatest easel in the world, but if you can't sell it, what happens to it? We fought constantly with cost." Developers, she feels, need to work with publishers to make their projects marketable and salable.

Electronic networking is a core feature of the Kids Network project. Cost of networking time for the field test sites is paid by NGS which had an interest in electronic publishing. It was that interest that stimulated the early talks between Monica Bradshaw of NGS and Robert Tinker. In later years, when speaking of this planning, Tinker said that they were all naive about costs which were greatly underestimated. Asked in an interview what he has learned and would do differently he said, " . . . in retrospect, we should have insisted that they [NGS] fund a person part time or full time from the very beginning to lay the foundation for the work that needed to be done later on . . . so they didn't always have to play catch up . . . and felt more part of the project."

TERC developers talked about how the content and the "science" suffered in the hands of NGS editors who often rewrote or dropped sections. Early and continuous collaboration in field testing and editing might have resulted in a product about which all parties would be more satisfied. Tinker described developer-publisher tensions as creatively useful, provided both parties make some appropriate accommodations.

As TERC and NGS have joined to move into a new middle school effort and revision of some of the elementary school materials, both organizations have profited by their knowledge of the other and what happens to their products in schools. New ground rules have been developed to reflect what they have learned about how to use the knowledge each has to get good science that is marketable into their project. But success will depend on long-term involvement of publisher staff familiar with the agreements—particularly in light of the fact that turnover of personnel in publishing houses is remarkably frequent.

Technology Issues

Since the time when most of the projects featured in our case studies began, the technological picture in schools has kept changing, often unpredictably. This has created both opportunities for developing innovative curricula and teaching strategies and hurdles for their use. Both these aspects of the increasing availability of educational technology are illustrated by our case studies.

Over the past decade, calculators, like computers, have become computationally and graphically more powerful. In the discussions associated with NCTM's development of its Standards documents, the role of graphing calculators was always addressed. Some people worried about public acceptance. Many parents feared calculators would undermine the development of good basic mathematical skills. In the writing teams, there was debate about whether graphing calculators ought to be available by the 11th grade. As NCTM collected feedback from the field about the *Curriculum and Evaluation Standards*, it became clear that sentiments were changing. If graphing calculators were good enough to serve the mathematics curriculum for 11th and 12th grades, why should they be kept out of the 9th and 10th grades?

Their graphing capabilities eventually made calculators an important adjunct to instruction in the *Contemporary Precalculus Through Applications* course. The instructional approach in this course was to begin with an applied problem and then to learn the mathematics needed to solve it. Computer spreadsheet software, powerful graphing calculators, and easily managed apparatus for measuring phenomena gave students firsthand experience in collecting and manipulating data, enabling them to learn skills in problem analysis and application of mathematical models to real data. The NCSSM teachers involved in developing the Precalculus course found the graphing calculators so useful, they created a separate calculator guide book to accompany the course. In addition, each of the teachers had a graphing calculator modified for use on an overhead projector.

In the UMC project, some mathematics teachers with strong interest in the use of technology saw that graphing calculators would allow them to introduce new mathematical topics that otherwise would be impractical to consider. They actively encouraged peers to use the calculators. Their participation in the UMC network allowed them to spread word of their experiences as well as to hear about the uses to which other teachers had put the graphing calculators in a variety of mathematics courses offered at different grade levels. Among UMC teachers, computers and calculators have come to be regarded as tools essential to doing modern mathematics, not just as useful pedagogical devices.

The addition of modems to more powerful computers has opened the door to low-cost electronic networking for users. But the opportunities afforded by the availability of a network may fail to be realized. The ACS ChemCom project developers, for example, sought to take advantage of electronic networking to establish a bulletin board for ChemCom teachers. After two years, ACS closed the bulletin board for lack of sufficient use to justify its expense. Yet, nearly five years later, two ChemCom teachers opened a bulletin board about the course; participation, while still small, is increasing. Teachers are sharing problems, getting and giving answers to questions, and making new curricular suggestions. How long the activity will continue remains to be seen. Although hardware capacity for school-based electronic networking is increasing, participation of science teachers generally is small, given their numbers. One problem seems to be the lack of available phone line access at schools.

CD-ROM players now are widely purchased in the home computing market and are becoming increasingly available in schools. Project 2061 has taken advantage of this technology to make its publication *Benchmarks for Science Literacy* (AAAS 1993) available on CD as well as in the standard book format. The developer designed an interface in such a way that users can move around the *Benchmarks* database quickly and flexibly to find what they need. The software design helps users explore how well a particular curriculum conforms to the *Benchmarks'* recommendations.

Videoplayers and television have replaced the once ubiquitous film and filmstrip projectors in schools. Once videoplayers became a widely used and familiar technology in homes, schools followed suit—but not very rapidly. To advance the use of technology for enriching the curriculum, the federal government's Office of Education (the predecessor of the U.S. Department of Education) put out a request for proposals (RFP) to create curricular products that exploited what were then new technologies: computers, videoplayers, and interactive videodisk machines. The RFP was quite specific about what was to be produced: 26 quarter hours of video, an interactive videodisk, and computer software. The program officer wanted to pilot interactive technologies.

Samuel Gibbon, who submitted the Mimi project proposal in response to the RFP, saw little evidence of videodisk players in schools. He suggested that Mimi make videotapes, which could be used interactively or not in conjunction with an Apple computer. Gibbon was betting that video rather than videodisk technology would make faster headway in schools. Nevertheless, his team did eventually produce a videodisk product. He remarked that, at the time, even asking teachers to use video tapes was problematic enough. Videodisk technology with its requirement for a player, a computer, and a television set all interacting was many degrees more costly and complex for users.

The Mimi case also presents an example of how the appearance of ever more powerful computers poses problems for developers as well as publishers. Mimi's first software development was done for an early generation of Apple computers which soon went out of date. As newer models with more capabilities—graphics, greater memory, greater computing power—are acquired by schools, projects and their publishers have to modernize or leave the field to their competitors. But paying for new software development or even for basic updates is a problem. Gibbon's solution was to switch to a different publisher for the second Mimi series, a multimedia publisher who supported conversion to the Macintosh platform.

The technology dilemmas posed for computer-based projects is further illustrated in the Kids Network project, as the comments of Robert Tinker, its originator, make clear:

> The history of the technology is that we developed it through the Apple IIgs. At that time, the IIgs was a brand-new computer, and it looked like the one that was going to take over the market. It had twice the resolution of the Apple II and a mouse which gave it MAC-like point-and-touch capability. It turned out of course that IIgs was abandoned largely by Apple. We complained loudly and got Apple to give us 200 IIgs [computers] for our early field tests . . .

> We're never funded to do software development directly. We're funded to do a curriculum project, then you have to bootleg the technology alongside it . . . So, there's never enough money to do the technology the way we want to do it. We . . . spent a lot of money that was not originally allocated to do the IIgs software.

The arrival of more powerful computers and faster modems made electronic networking an increasingly attractive opportunity for expanding curricular horizons. But when Tinker had to convert the already complex software to other and newer platforms, he had a problem. There was no money left in the grant to do it. IBM helped him put the Kids Network software on its platform. Then it was up to the publisher, NGS, to support conversion to the Apple Macintosh platform. As Tinker remarked:

> The software is outrageously expensive. We're talking about a word processor, a data analysis package [with graphics] . . . hundreds and hundreds of thousands of dollars which was in nobody's budget . . . We were really way ahead of anybody else in simplifying this telecommunications and data-sharing analysis software. That's the good part. The bad part is that we never had the resources to do it right.

The world of modems which makes networking so accessible also has been advancing technologically. Each new generation of hardware can transmit more data more rapidly. Taking advantage of the more powerful computers, developers with means can redesign their network interfaces to make them easier for users to operate. The Kids Network interface software continues to undergo changes.

The pace of technological change will probably increase over the next decade. What this will mean to developers and publishers, to say nothing of users, remains to be seen. From a policy perspective, it would seem that agencies and foundations that fund technology-dependent curriculum projects need to build into their budgets provisions for proper programming during development as well as for periodic review and updating of the software. New computers with enhanced capabilities require more than just an update. To exploit their potential, developers and/or publishers may need to redesign some of their software completely. That can be an expensive undertaking. It may well be that support for software conversions and updates ought to be the responsibility of the publisher of a project, once a project is on the market. At the very least, issues of technology updating should be taken into consideration in negotiations involving funders, developers, and publishers.

6

Underplayed Issues

Robert E. Stake
Senta A. Raizen

In previous chapters, we have presented the major common features of the eight mathematics and science educational reform efforts we studied. Next, we consider some issues that might well have been given greater attention by the reformers—and hence the case study researchers—but were not, perhaps surprisingly in view of the current context of reform described in chapter 2. These underplayed elements were the assessment of student achievement; formal evaluation of the reform efforts; and issues of equity, inclusion, and diversity of student populations.

Assessment of Student Achievement

As noted in chapter 2, critics of American education have repeatedly pointed to poor student performance in science and mathematics. As measured by standardized tests, for more than two decades the academic performance of American students has been seen as distressingly low. But the innovations we studied have been uninterested in direct methods of improving such test scores. Rather, they have concentrated on the content and pedagogy of science and mathematics classrooms. Implicitly, the reformers responsible for these innovations either did not concern themselves with the widespread belief in the appropriateness of using conventional standardized tests to indicate the quality of student performance in science and mathematics, or, as a result of active consideration, they found such tests inadequate for assessing the kinds of ambitious student learning they envisaged. Hence, as documented in the eight case studies, the focus with respect to assessing learning was much more on academic integrity than on expressed parental or public need for traditional test results. As a group, these reform projects—with the exception of the California reform—did not invest much effort in upgrading assessment information for persons other than teachers.

What Kind of Assessment?

Though they did not include the goal of improving standardized test scores, the comprehensive projects and several of the curriculum development projects raised the issue of assessment of student achievement. The main message was that assessment practices also needed to be reformed as part of the logic and util-ity of the new teaching. Yet it is difficult to assure parents, taxpayers, and skepti-cal teachers that the new curricula and teaching strategies will provide students the information that achievement testing has traditionally required. Reformers who claimed that back in the 1960s failed to be persuasive (Stake and Easley 1978).

Evaluating Program Effectiveness. When a new approach to teaching or a redefinition of curriculum comes along, most people expect it to be evaluat-ed in terms of the best measures of student achievement already available. Yet almost all teaching can be expected to look ineffective if the criterion is *change in scholastic aptitude.* This is precisely what psychometricians have devel-oped—a highly refined system of measuring students' scholastic aptitudes. And since, in a heterogeneous student population, scholastic aptitude correlates high-ly with teacher grades and other measures of achievement, aptitude items that look like achievement items have been used throughout the country to assess levels of achievement, even though technically they do not measure achieve-ment (Stake 1995).

Innovative programs will be particularly vulnerable when such tests are used to judge program effectiveness. They are certain to appear ineffective, especially if they do not organize some of their instruction around this type of criterion testing. It is possible and desirable for them to create some of their own criterion-based examinations, as happened in the California reform, but indepen-dent outsiders as well as opponents are going to presume that the reformers' tests will be biased toward their innovations. Conventional examinations will be treated as an unbiased crucible. Thus, an innovation needs—as pointed out in the California case study—not only to reconceptualize teaching and learning but assessment as well.

Improving Instruction Through Assessment. Political rhetoric to the contrary, conventional tests have been of little use for diagnostic purposes, sel-dom indicating what remediation was appropriate. It is possible that the tests could be better used to improve instruction if practitioners were better educated in diagnostics. We think it more likely, however, that information for improving instruction cannot be found in conventional test data because diagnostic infor-mation is not inherent in the elicited student responses. Worse still for diagnos-tics, an educational technology relating test performance to needed changes in teaching is not yet available, even after decades of trying. (Some of the most dedicated and ingenious educational researchers have tried, without much suc-cess, to develop the underlying theory. See Cronbach 1977, and Snow and Mandinach 1991.) Nevertheless, as discussed later in this chapter, France has instituted a large-scale diagnostic assessment in mathematics.

Assessing Student Learning Through Performance Tasks. Current reform recommendations are supportive of "performance testing," that is, testing that places the student in a problem situation where strategy and multiple sources of information are required, where contexts are somewhat realistic, and where students have the opportunity to work on solutions (sometimes in groups) that take a while. An example of such a performance task is provided in *California's Science Framework* (California Department of Education 1990, p. 51): " . . . design an experiment that would test the effects of variable temperature on the respiration rate of goldfish." Such a task takes a lot of testing time. This creates two problems, not so much because the task or item is unconventional, but because only a few such items can be included in an assessment. The problems are inability to probe adequately students' knowledge of a given content domain, and test results that are sometimes untrustworthy and difficult to generalize.[1] Indeed, a move toward substituting performance testing for conventional testing gained headway in California (see "Assessment in the Comprehensive Projects") but ultimately was frustrated. Other than in the California case, the case researchers did not report confrontation of state-mandated testing in the innovative science and mathematics programs they were studying. Opposition to the demand for norm-referenced, standardized testing may be too costly a battle for most curriculum reformers.

To many teachers, now as in the past, teaching and subject matter integrity have much more to do with what students do in the classroom than with what they do on tests. Whether the emphasis has been on "sit and git" or on "action and interaction," the obligation of the teacher—as conceived by teachers and many other educators—has been to make the *experience* of education what it ought to be. Teachers have seldom expected external tests to capture what they were trying to do. Yet most teachers have come to accept standardized tests as important indicators of student progress and to see that preparation for those tests is one of their obligations. At the same time, innovative teachers have come to see performance tasks as an essential part of instruction, useful even if their inclusion in large-scale assessments and reporting of results remain problematic. Teachers and other educators committed to performance testing but also needing norm-referenced information will have to find ways to extend the testing period.

Aligning Instruction and Assessment. There is agreement within the education and evaluation communities, whether reform-oriented or not, that there should be alignment between student assessment and instruction, that the best of the teaching should be recognizable in the testing. It could be argued that good assessment also should point out what has not been taught that could or should have been taught. However, since there is no national curriculum taught

---

[1] Whenever assessment is based on a small number of items—in this case, because they take so long—the results are likely to be psychometrically unreliable. With so few answers to produce, sometimes a poor student stumbles onto a high-scoring answer; sometimes a good student just stumbles. Occasional deviation from usual levels of response balances out when the set of independent items is large. Moreover, Shavelson and Baxter (1991) have shown that it cannot be concluded from a student's performance on a small number of tasks how he or she might perform on other seemingly similar tasks. One problem is that the cognitive demands of a complex task are difficult to identify, let alone duplicate in other tasks.

in the same way in every classroom, the assessment could be unfair if it concentrates on what students have the opportunity to learn in some schools and omits what they are expected to learn in others.[2] Although schools already find that testing takes too much time, one could envisage comprehensive batteries that would cover the various expectations. But under long-standing scoring rules that count all items equal, the fairness of even broader assessments is problematic, the argument again being that it is not fair to grade students on what they were not taught. Of course, that raises the issue of the purpose of testing: Is it to learn what the students know or what they have been taught? Both questions are legitimate, and one of the shortcomings of the assessment movement has been the failure to differentiate well between the two. Among policy setters within the public and the profession, we find quite divergent expectations of students. Their views should not be disregarded. No one view should be taken as the only basis for designing the assessment of student progress.

Assessment in the Innovations

The view of the reformers involved in the eight innovations we studied has been somewhat conflicted as to who should design assessments. As pointed out in the case studies of the California reform, the Urban Mathematics Collaborative (UMC), and several of the curriculum development projects, the reform efforts wanted to draw greatly on the wisdom and experience of teachers. The content and pedagogy were to be considerably determined by teachers (and students too). Teachers were to be the authorities on teaching, without being authoritarian. Yet reformers wanted teachers who believed conventional tests to be important indicators of student progress to back away from that belief. It has been a difficult dilemma. Experience has shown that teachers can only rarely be counted on to insist that external testing be aligned with their most important curricular efforts.

Assessment in the Curriculum Development Projects. Of the four projects in this category, ChemCom invested the greatest effort in developing an assessment of student learning based on its innovative curricular approach and content. A special nationally normed, machine-scorable ChemCom test was developed, to be taken over two class periods. In addition, the teacher's guide for the ChemCom textbook contains tests to be given at the end of each unit, although the case study notes that teachers seemed to prefer writing their own

---

[2]There now are nationally developed standards available in both science—including *Benchmarks* (AAAS 1993) and those published by the National Research Council (1996)—and mathematics—the National Council of Teachers of Mathematics *Standards*. Two of our case studies provide further detail. There is, however, far from universal agreement on any of the standards documents, either in science or in mathematics. Moreover, none of the standards documents is spelled out in sufficient detail to ensure uniform curricula in all classrooms, nor is that their intent. Translation into specific curricula, teaching strategies, and assessments is left to individual states, districts, schools, and teachers.

tests. Teachers experienced some grading difficulties because of the unconventional nature of the course, particularly the segments that asked students to debate a given issue on the basis of their chemistry knowledge—for example, the pros and cons of aspirin use (see Rowe et al. 1996, pp. 568-70).

The Precalculus group found that it had to develop assessment materials to accompany the textbook it had been encouraged to write. The assessment included complex applied problems that involved realistic situations, opportunities for exploration, and several mathematical concepts. The example given in the case study involves having to distinguish between two types of midges— important because one type is a human disease vector—on the basis of data about wing length and antenna length (Kilpatrick et al. 1996, pp. 157-58). Individual teachers using the course also addressed assessments independently; for example, a teacher at one of the adopting schools used independent student projects, group work, and individual written and oral reports to evaluate students' learning; students also could get extra credit if they kept a daily journal on their reactions to the mathematics classes.

In Kids Network, student achievement testing was given little priority. The project materials distributed by the National Geographic Society to support the curriculum units did not include specific assessment materials for teachers' use.[3] According to the case study (Karlan et al. 1996, p. 282), teachers "reported that they made informal assessments around students' levels of engagement and not around specific content knowledge . . . " Teachers appeared more interested in students' acquiring science process skills than understanding scientific concepts.

As for Voyage of the Mimi, Gibbon (the executive director of the project) is quoted in the case study (Middlebrooks et al. 1996, p. 411) as saying:

> We got from the publishers very strong pressure to put paper-and-pencil tests in all the material. Of course, their other materials did that. We saw this as perverting the Mimi materials. All that the episodes and expeditions programs offered would be contravened by tests that said, "These are the important and, therefore, the only things you ought to teach to . . . " Conventional testing does not fit with Mimi.

Indeed, there is no formal assessment component in the Mimi materials, including those intended for teachers. The case study researchers speculate that the absence of the assessment component may be a factor in the limited use of the computer materials. Most teachers, however, need to give grades in science. According to the case study, teachers reported relying for this purpose on traditional subjective methods: enthusiasm and amount of student participation, quality of oral discussion, and completeness of written assignments.

Assessment in the Comprehensive Projects. The National Council of Teachers of Mathematics (NCTM) standards effort envisaged from the outset

---

[3]The Kids Network materials are currently undergoing revision. One of the express purposes of the revision is to develop appropriate assessment tasks for teachers' use.

that a series of documents would be produced, one of which would address assessment. *Assessment Standards for School Mathematics* (NCTM 1995) appeared six years after the original standards document addressing curriculum and evaluation. Even that earlier document (NCTM 1989), however, had included in its evaluation section a discussion of desired attributes of student assessment—alignment with the curriculum, multiple sources of information, appropriate assessment methods—as well as examples of assessment tasks related to the curriculum standards discussed in the rest of the document. The later document, devoted wholly to assessment standards, lays out additional assessment standards dealing, for example, with making instructional decisions and evaluating the mathematics program. (Is it coherent? Are all students learning?) It also provides detailed standards for assessing student learning, both to monitor students' progress and to evaluate their achievement. Together with general guidelines on what changes in assessment practice are needed, the document gives additional examples of tasks appropriate to gauge mathematics learning based on an NCTM standards-linked curriculum. The assessment document recognizes the need for reporting student achievement to the public, but urges that this not be based on a single letter grade or number; descriptive summary reports and work samples are suggested instead.

In the other comprehensive mathematics case, the Urban Mathematics Collaborative, the case study researchers (Webb et al. 1996, p. 299) found that: "Many collaborative teachers questioned the fairness of standardized tests, which did not measure the same outcomes they viewed as important for students in mathematics, as well as the usefulness of standardized tests as formative evaluation tools for themselves and their mathematics programs."

Instead, the collaboratives offered collegial exchanges, networking among teachers, and keeping teacher journals accompanied by regular discussion meetings within a school as ways to increase teacher reflection and improvement of practice. Some teachers, encouraged by their technical assistance provider, experimented with a variety of "authentic" assessments—student portfolios, student journals, and using realistic contexts for mathematical problems, made possible by the use of the calculator. Other teachers, however, felt they had to continue preparing their students for the standardized tests mandated in their districts.

Although the focus of Project 2061 has always been on the *science content* to be taught in grades K-12, early on the project also laid plans for providing implementation guidance. From its very initiation, a series of "Blueprints for Reform" were planned, to be written by eminent authorities in the field. One of these was to be on assessment. It was to "specify what immediate and future assessment needs are demanded by Project 2061 curriculum-design principles, from in-class assessment during instruction, to program evaluation by schools, to monitoring educational progress at state and national levels" (AAAS 1995, p. 23). At the time of the case study, drafts of the assessment blueprint had been written but were not available for public distribution. It is difficult to see what the project might have done more directly to assess student learning in science programs based on Project 2061, since such programs are still far from implementation in the schools.

The California reform, as noted earlier, is the only one of our eight cases to put concerted effort into the reform of large-scale assessment. The *California Science Framework* itself indicated that assessment of student learning was very important, championed good assessment, but did not deal directly with its provisions. The state already had an elaborate centralized student assessment system, as sophisticated as any in the country and more sophisticated than most. It was based on the conventional idea that each student should be tested on a uniform body of knowledge and skills so that scores could indicate the relative standing of students, rather than their being tested to inventory each student's actual knowledge and skill.

At the secondary school level, as the case study indicates, the issue of reforming assessment was taken up by the state-funded Scope, Sequence, and Coordination project. Model assessment items for possible teacher use were generated by a group of participating teachers. Operating separately along traditional lines of high-stakes, state-mandated assessments, specialists in large-scale assessment (drawn together by the State Department of Public Instruction) designed a battery of tests called CLAS—*C*alifornia *L*earning *A*ssessment *S*ystem. The tests were a combination of multiple choice items, justified multiple choice items (where students had to give reasons for their responses), open-ended items, and hands-on investigations.

But, following the lead of the state's governor, the California Legislature did not provide funds for CLAS administration, partly because of a perception that the CLAS tests were not sufficiently strong in addressing basic scientific knowledge, introducing instead methodologically unproven approaches such as performance tasks. The loss of CLAS cut short a valuable effort to seek compromise between standardized testing and the state's forward-looking, innovative framework—what might have been part of the greater accord being sought between traditional and reform approaches to the science curriculum and instruction.

Among the design principles for CLAS were (Atkin et al. 1996c, p. 92):

1. The assessment should model good instruction.

2. Assessment should be coordinated with the curriculum and viewed as an instructional task rather than as an isolated opportunity for evaluation.

3. Assessments will have no single prescribed answer, but will allow for a variety of appropriate responses.

4. Teachers will be heavily involved in the development, piloting, field testing, administration, and scoring of the CLAS tests.

These principles envisage assessment as a facilitator of instruction. They are not the principles for norm referencing, nor the principles for the generation of criterion instruments to ascertain that a student is qualified, that a teacher is teaching well, or that a state educational system is in proper repair. The CLAS principles stem from legitimate evaluative concerns and questions, but the technology for responding to them is far weaker than the technology for ranking students, classrooms, or districts. It is not clear whether the rationale for CLAS included an expectation that normative information would accompany CLAS scores, would go unsought, or whether conventional testing would be used to answer them.

In summary, most reformers in the eight projects we studied agreed that the reconceptualization of science education is incomplete if it leaves out the reconceptualization of assessment. Yet systemic educational reform calls for the use of rigorous, objectively scored, standardized tests as bottom-line criteria. Society has a huge and legitimate stake in reform. It has to choose criteria meaningful to it. In the end, reformers may have to compromise with the American confidence in externally mandated, standardized testing not under the control of the teacher. This challenge has not been resolved either in these projects or within the rest of the science education or evaluation professions in the United States.

Assessment Approaches in Other Countries—Testing to Help Learning

Two of the other 12 countries participating in the Organisation for Economic Co-operation and Development (OECD) case study project selected for study an innovation in assessment. These countries, France and Norway, went considerably farther in reconceptualizing assessment than did any of the U.S. projects. Brief descriptions of the case studies by France and Norway are provided in the appendix. Below, we summarize salient findings from the case study on each.

France. When France, between 1989 and 1992, instituted its all-student testing program in mathematics for grades 3, 6, and 10, the intent was to provide teachers with a means for evaluating the readiness of their students for new material and shape any necessary remedial instruction accordingly. In short, the purpose of the assessment is as a diagnostic service to teachers; hence, the tests are administered at the *beginning* of the school year, not at the end, as is the usual practice.

Moreover, the information sent back to teachers provides more than just scores on individual items. So as to help teachers in adjusting instruction, the tests probe not only mathematical knowledge but also students' ability to engage in thinking processes and perform the cognitive operations of mathematics. Obviously, the accuracy and relevance of teachers' diagnosis of their students' mathematical development is critical to their ability to shape their instruction effectively. Hence, to help teachers interpret testing results, test questions are classified according to two dimensions: mathematical content area and intellectual operation (including skills and techniques). Two types of booklets are prepared for each assessment: One type is the test booklet for the students, the other explains to teachers how to analyze individual students' responses according to the dimensions represented by each question. Recording the results on each question for every student in a class generates quite a large database, so teachers also are provided with software for processing the student data.

During the three years that the assessment was introduced, the education ministry provided intensive teacher training. The training focused on the interpretation of test results and appropriate ways of providing remedial instruction for the students needing it. Both teacher instructors and teachers also prepared brochures on their most effective teaching strategies and shared them with their peers. While training relevant to the assessment continues, it no longer constitutes the principal theme but is integrated into France's ongoing program of teacher training.

The case study of this assessment innovation found that teachers liked the approach, especially at the two earlier grade levels. Some noted that they found certain of the student responses very revealing; they helped teachers appreciate the level of mathematical understanding of their class and deal with the differences among their students, for example their mastery—or lack thereof—of decimals. Teachers also said that they used the test results as a basis for their discussions with parents about their child's performance. Reactions varied somewhat by mathematical subject matter: Teachers noted that there were too few challenging questions dealing with number concepts but too many difficult geometry questions. Another concern expressed by teachers was the burden imposed by the annual repetition of the assessment.

Teachers used both the student test booklet and the teacher booklet to inform their teaching. Exercises in these materials were taken as models to broaden the scope of instruction. It seems that for many teachers, this aspect was the most useful part of the assessment, since they considered the test data only a partial reflection of students' knowledge, and one that would soon change as the year proceeded.

The case study researchers conclude their report with the following points:

- Continuing training is necessary to ensure that all teachers, not only the highly motivated ones, will make effective use of the information provided on each student as well as the class as a whole.

- Further research is needed on the different cognitive demands posed by the questions—for example, through the analysis of errors—so as to strengthen the theoretical underpinnings of the assessment and suggested teaching strategies.

- As it is difficult to move teachers from practices based on their experience to practices grounded in research, trainers of teachers need to be competent not only in their subject but familiar with research on teaching and learning.

Norway. Norway has no formal assessment of student learning in the first six grades; in grades 7 and 8, teachers give numerical grades two or three times a year in the compulsory subjects, including mathematics. At grade 9, however, written examinations (prepared by a national board) and oral examinations (prepared by local school authorities) are given. Norwegian researchers found that both teachers and students experienced difficulties with the mathematics curriculum, particularly in grades 7-9. Teachers thought there was too much to cover to prepare students for the ninth grade external written exam. Students found the subject difficult, with too much to memorize; they also didn't see much use for what they were learning, although they thought they might need it later.

The main focus of Norway's assessment innovation is to develop students' ability to assess their own mathematics learning, with two purposes in mind: to make assessment function in support of instruction, and to make assessment more meaningful to students. The project leaders developed some self-assessment methods they thought suitable for seventh- and eighth-graders, the grades targeted for the innovation. The materials included new types of test questions, student-written personal logs, portfolios students could keep of their work, and self-assessment sheets to be filled out in connection with tests and also—more generally—with respect to their mathematics learning. The intent was that students would see assessment activities as part of their mathematics learning. Participating schools were free to use these materials as they thought best. Some schools added further approaches, for example, having students select their own tasks, including some thought up by the students themselves, or doing tasks in groups.

Not all the materials proved equally popular. Log writing was perhaps the most difficult activity, particularly for slower learners—although in schools that introduced them systematically (providing feedback, for example, and not overusing them), logs were thought helpful by about 40 percent of the students. Self-assessment sheets were found variously useful by different classes, depending somewhat on the use being made of them. Portfolios at first caused some anxiety for students, but at the end of the term, students commented that they were proud of their work and gained a picture of their standing in mathematics. The national examination board and participating schools currently are considering how to make portfolios part of the ninth grade oral exam.

The case study researchers concluded that teachers had changed both their teaching and their assessment methods as a result of the innovation. They now realize the importance of openness, frequent and rapid feedback, and supportive correction. As one teacher commented (Jernquest 1995, p. 24): "Involving the . . . pupils strengthens the effect of assessment in the learning process. Pupil self-assessment has as a consequence that they are far more motivated and conscious in relation to their work. They are more responsible, and their efforts are more long-term and goal-centered."

Evaluation of the Innovation Itself

Program evaluation may well include assessment of student learning, but it generally needs to go far beyond a single type of outcome measure to establish the effectiveness of a program's design, administration, and impact. With the current emphasis on accountability, considerable attention might have been expected of reform projects in evaluating their programs. In a sense, the case studies conducted by several of the countries participating in the OECD project were evaluations of their innovations in science, mathematics, or technology education. This is not surprising since, in most of these countries, governmental authorities had both initiated the reform *and* selected it for study. For example, in Spain, the education ministry mandated two reforms at the same time: raising compulsory school-leaving age from 14 to 16, and making science for ages 12-15 an integrated subject. The science reform was to be introduced over an eight-year period, being voluntary during the time of the case study. The ministry was keenly interested in how the science reform in the volunteering schools was faring, namely, how teachers were implementing the new approach to science, which previously had been taught as separate subjects and—at ages 15 and 16—only to a select population. As another example, Ireland found that the low enrollment of girls in physics and chemistry was due to their lack of opportunity to study these subjects in their schools. The ministry of education mounted an in-school teacher development effort, supported by expert teachers from other schools, curriculum materials, and ongoing teacher centers. The Irish case study was undertaken because the ministry wanted to know to what extent this effort had succeeded in increasing enrollment of girls in the physical sciences. Similarly, the education ministry in Japan was interested in finding out how teachers were handling the new curricula in science and mathematics introduced in 1992 and 1993.

In contrast, in the United States, neither the funders nor the leaders of the eight innovations asked for the case studies, although all of the latter cooperated fully with the case study researchers. Nor would we hold that our studies represent evaluations of the projects, limited as the case studies were to a year's investigation. This is particularly true of the comprehensive projects, which by their very nature envisage many years of implementation effort. It is appropriate to ask, however, whether the projects themselves commissioned external evaluations. The case study researchers cite little evidence to that effect. It appears that the reformers showed limited interest in obtaining program evaluation studies to document changes in the content and quality of science and mathematics education.

Program Evaluation in the Curriculum Development Projects

As is common in the development of innovative curricula, the four curriculum projects all tried out pilot versions of their materials and used the resulting feedback to improve them. Small in size and minimally funded, the group responsible for the Precalculus project did this type of evaluation informally, as individual teachers compared their experiences with specific units. The developers of ChemCom, in addition to trialing draft versions of each unit, systematically evaluated their teacher training institutes. Some independent researchers doing small-scale studies found that ChemCom students did as well or better than students enrolled in traditional courses on tests of chemical knowledge, and better on items assessing application of chemical knowledge and making related decisions (Rowe et al. 1996, pp. 571-72). The sponsoring American Chemical Society itself, however, never commissioned any external evaluation of ChemCom.

The case studies of Kids Network and Voyage of the Mimi concentrated on enactment at school sites, paying but minor attention to the development process at Technical Education Research Centers and Bank Street College, respectively. Formal field-testing of prototypes and a formal external formative evaluation were conducted for Mimi, with key views drawn from participating teachers. These evaluations led to several changes in the materials, particularly the computer simulations. There is no mention in the case study, however, of any external evaluation of the quality and effectiveness of the Mimi project as a whole, nor is there any such mention in the Kids Network case study.

Program Evaluation in the Comprehensive Projects

Because of the scope of these projects, we discuss documentation and evaluation for each project individually.

The Urban Mathematics Collaborative. The Ford Foundation, which funded this project, commissioned as well an external study to document its processes and progress for the first five years. The resulting monograph reports a number of findings (Webb and Romberg 1994, pp. 6-7): flexibility in design was essential so that the collaboratives could take different forms appropriate to each participating district; establishing effective communication between business executives and educators was time consuming and sometimes difficult; nevertheless, teachers were enthusiastic about their involvement in the collaborative and were recognized in their districts for their professional contributions. The documentation report also comments that the problems of urban teaching were not alleviated nor even addressed. That, however, was not the objective of the Ford Foundation, as the report points out; rather, the Foundation was interested in the empowerment and effectiveness of individual teachers. In the eyes of the researchers who conducted the subsequent case study, this documentation effort did not constitute any sort of evaluation. In fact, they comment (Webb et al. 1996, p. 347): "Among the 16 collaboratives, systematic self-evaluation of programs, administration, and impact were notably lacking" despite the need expressed by local administrators for such evaluations. And while the outside organization providing technical assistance to the collaboratives raised a number of evaluative questions, these concentrated on student learning rather than program evaluation.

The NCTM Standards. McLeod et al. (1996) discuss the notable influence of the *Standards* on the shape of the mathematics and science education reform movement and on state documents, teachers, and other communities, as documented by a variety of surveys and studies. In addition, a number of independent studies have addressed the question of impact of the *Standards* at the classroom level (see, for example, Porter et al. 1993). NCTM itself early on set up a task force to monitor the effects of the standards. This resulted, among other activities, in a project funded by the Exxon Education Foundation to identify and describe examples of changed practice consistent with the standards (Ferrini-Mundy and Johnson 1994). The case study researchers, considering a variety of studies and on the basis of their site visits to several schools, provide their own view:

> NCTM as an organization is clearly concerned; in 1995 a new task force on Professional Outreach . . . was trying to address forces that oppose the *Standards* . . . reform in schools continues, and there are notable signs of progress . . . But change occurs slowly, and the public shows little inclination to provide the sustained effort and financial support that significant change would require . . . NCTM has made little progress toward promoting a new conception of mathematics; too few teachers and parents see understanding mathematics as the goal, and too many are still satisfied to focus on rules and procedures without understanding (McLeod et al. 1996, p. 121).

Project 2061. Starting in 1993, Project 2061 undertook a sizable documentation effort to collect self-descriptions of field site activities and impact. According to the Project 2061 staff, the documentation process grew out of the interest of the teachers at the individual sites to capture what they were learning. The process also helped inform the project of its progress and at the same time build credibility for future funding (Atkin et al. 1996b, p. 148). To serve these purposes, the reports coming in from the sites were compiled and analyzed centrally. The documentation reports were not intended as "official" reports on the activities and progress of the sites. Their main purpose was to help the teachers reflect upon their work in curriculum development, dissemination, and implementation. Hence, even though an outside researcher was involved in guiding the documentation activity, it was not primarily a program evaluation effort in intent, in design, in the information collected, or in the use made of the reports from the sites. As the case study was coming to a close, Project 2061 commissioned SRI International to conduct an external evaluation of its activities and impact.

The California Reform. The *California Science Framework* was the basic component of the California reform effort. Although it underwent several review cycles, formal program evaluation to assess its progress in the schools was not built in. By the time the *Framework* was five years old, its primary implementation mechanism, the California Science Implementation Network (CSIN), though demonstrably engaged in looking at its problems and strengths internally, had not engaged a critical independent body to examine the quality of the work. According to the Atkin et al. (1996c) case study, CSIN's primary criterion at this stage was to provide teachers with educational and professional experiences that would engage them in the long-term process of change in elementary school science. But there was no effort to assess these experiences formally, especially whether they had any effect on the rank-and-file teachers who were not a direct part of CSIN. Professional development generally suffers the "scaling-up" problem—that is, getting the benefits of the experience spread to those teachers who do not have (or who, if they do, resist) direct opportunity for participation. Though the California reformers held explicit views of how their innovations would spread, the case study report does not indicate any specific initiative to evaluate the scaling-up effect in California science education.

Nevertheless, there was wide realization by the reformers of the need for and difficulty of formal evaluation of the progress of their efforts. (One of the reform leaders was herself an evaluation expert.) Awareness was stimulated particularly by the National Science Foundation (NSF) and other funding agencies calling for indicators or benchmarks of progress. The reformers, however, knew that "objective" indicators of progress collected over the relatively short life of the reform were likely to be invalid, if not harmful. Attempting to provide assurances of success in terms of student test scores—as demanded by some—might well jeopardize the whole effort. Assessment of student learning must be part of any curriculum, but program evaluation cannot be limited only to the criterion of increase in standardized, normed test scores. Clearly, a proper evaluation needs to be sensitive to what reformers are trying to accomplish, but also to the requests of funders and policymakers for information on the effectiveness of the reform. This is a difficult balance to attain in program evaluation. So perhaps it is not surprising that those driving the California effort did not—in the same creative and conceptually sound way they set up change mechanisms in content and pedagogy—set up a mechanism for disciplined evaluation of the reform. They were amply attentive to informal indicators of progress (and regress) but they did not draw upon experts in policy study and evaluative field study to provide independent assessment of their own strategies and activities. They exhibited the usual reluctance to invoke formal evaluation—due partly to its cost and partly to the fear that the results would be more hurtful than helpful.

Program Evaluation as Part of a Rational Change Strategy

Although it was not reported in the case study, another reason that no formal evaluation was commissioned in California may have been doubt that evaluators would really appreciate what the reform was trying to accomplish. The experience of most curriculum reforms seems to be that evaluation studies do not get to the heart of the matter (Berlak et al. 1992). They do not provide definitive answers as to the quality of the lesson materials or teaching. The scientists and mathematicians involved in the 1960s reforms had gone looking for keen-eyed researchers to gather empirical evidence of teaching quality. For example, the major mathematics reform project of that era, the School Mathematics Study Group, launched the National Longitudinal Study of Mathematical Ability to track the effects of the reform on student learning in mathematics. But the success of those early evaluators had been modest—not enough to persuade this generation of reformers to make similar investments. For whatever reason, drawing upon the technological refinements of program evaluation was not prominent in the work of the innovation projects we studied.

The behavior of earlier and later reformers generally, especially as described in our eight case studies, is closer to that of creative artists than to architects and engineers. Measurement is not a trusted ally. There is greater confidence in the innovators' own capacity for perceiving quality and effect. Beyond a superficial level, there is doubt about formative data as indicative of project well-being. If innovators want a check upon logic, appreciation of vision, confirmation of problems, and suggestion for damage control, they go to trusted peers and fellow reformers—not to testers, auditors, efficiency experts, or program evaluators. The widespread requirement for evaluation is considered to be an exercise in bureaucratic accountability, not one in discovering the essence of what the reform is about. Though many of the reformers have in fact participated in evaluative studies, independent program evaluation has not emerged as central to the culture of curriculum and teaching reform in mathematics and science education.

Rather than relying on evaluative information to win the public to their reconceptualizations, the projects' promotional materials emphasize the modernity of the world the students are maturing into, calling for science and mathematics skills and understandings for all and readiness for further change—abilities and attitudes not taught by curricula of the past. The reformers interviewed in the case studies were well-aware of the public's skepticism regarding the general quality of education and certainly shared this view with respect to science and mathematics education. However, whether teachers, scientists and mathematicians, or science and mathematics educators, the individuals leading the projects strongly believed in their professional ability to bring about improvements rather than having them mandated by legislatures, governors, or judges as had become more and more the case in other areas of education (Sarason 1971, Darling-Hammond 1990). The logic informing the development of the individual projects was, for the most part, quite intuitive, choosing to build on particular knowledge and experience—the need to involve teachers, the need to address many components of the system simultaneously, the need for new dissemination mechanisms. It is not that the projects were irrational in conception; rather that they relied little on rationalistic change processes, capitalizing instead on social and personal connections to get things started and to keep them alive. Whether explicitly espoused or not, there is wisdom in the intuitive approach, a wisdom occasionally recognized in policy and leadership literatures. Although it is possible to superimpose *post hoc* a rationality on the manifestations of change, the change processes the case study researchers observed in the eight innovations were driven not by conscious or learned organizational techniques—for example, in assessment and evaluation—but by intuitive development of situational circumstance.

Issues of Equity and Diversity

The development of the eight innovations took place during a time of deep concern about the disparities of opportunity in science and mathematics education for different groups of students, with specific concerns about gender, ethnicity, poverty, language, and disability. For example, NSF publishes a report biennially on the status of women, minorities, and persons with disabilities in science and engineering. When Project 2061 chose as one of its slogans "Science for All Americans" and widely distributed a major document under that title, it expressed this concern for inclusion. It recognized that the facilities for science education are inequitable, with poor rural and urban and ethnic communities unable to provide for their children the science learning opportunities common in affluent enclaves. It was also expressing a concern for producing new generations of scientifically literate citizen, who would understand the science needed for personal and societal well-being in a modern world. Similarly motivated, the Mathematical Sciences Education Board (1989) published a report—closely linked to NCTM's standards effort—entitled *Everybody Counts*; the cover depicts students of all ages and hues, and includes one in a wheelchair.

How did the projects incorporate these equity concerns in their materials and activities? One would expect a strong emphasis on them in the more comprehensive efforts and a somewhat varied approach in the curriculum development projects. Therefore, we discuss these two sets of projects separately.

The Curriculum Development Projects

Obviously, the North Carolina Precalculus group did not set out to address the problem of increasing access to mathematics for currently underserved student populations. They were, in fact, dealing with a special population themselves, but not one that fits the standard categories. Their students were specially privileged, identified as particularly science- and mathematics-able. Because the developers were members, "elite" teacher networks proved to be a highly effective, if informal, dissemination mechanism to other privileged schools. Moreover, because of textbook adoption practices that tend to exclude innovative materials, teachers in many states and school districts did not have the option of using *Contemporary Precalculus Through Applications*.

From the outset, ChemCom was designed to appeal to students not attracted to the traditional 11th-grade chemistry course. The assumption was that an issues-oriented course that would introduce chemistry topics on an as-needed basis would prove relevant to many more students. The case study provides illustrations of the use of ChemCom in different settings, including rural high schools and high schools serving poor students and ethnic minority students. Because ChemCom focuses on applications and makes less mathematical demands than the traditional chemistry course, it is thought by many school administrators to be suitable to the greater diversity of students now enrolling in a second year of high school science—often chemistry. An unintended consequence is that, in some—though by no means all—school districts, ChemCom has become the course for the less able students. As the case study notes (Rowe et al. 1996, p. 560):

> There are many students enrolling in ChemCom who are at the lower academic end of their cohort. One teacher explains that his typical ChemCom student is in the bottom 40 percent of the graduating class and that a few of his students are at the very bottom . . . In tracked schools, [students] took it simply because they needed another science course for college, and ChemCom was the lowest ranked—and presumably the easiest—that was available.

The originators of ChemCom were interested in attracting more students to the study of chemistry, but not the less able ones. Their view was that ChemCom was as rigorous as the traditional course, although in different ways. Nevertheless, some schools and teachers used the course to provide opportunity for science study to students formerly excluded (or excluding themselves).

Both Kids Network and Voyage of the Mimi require a considerable investment from schools for materials and maintenance. The consequences were apparent at an early stage of the case studies, when the research teams selecting their sample schools for observation noted an underrepresentation of schools serving inner-city and high-poverty student populations. The Kids Network case study team found the developers sensitive to the problem. In fact, the developers in their piloting did include schools serving ethnically diverse populations and located in different areas—urban, suburban, and rural. The Kids Network materials try to reflect diversity by portraying people of different ages and ethnicities. The case study notes that the program appeals to students of different abilities; this view is reenforced by teachers who see the activities and the stress on group work not only as engaging to a wide range of students but also enabling all of them to learn. Yet the developers and disseminators were unable to sustain Kids Network in poor schools because of the annual subscription and upgrading costs. It was the exceptional principal who, like one in rural Vermont, made the program accessible to all teachers because: "It was an equity thing" (Karlan et al. 1996, p. 373).

The Mimi materials give special attention to age, gender, race, ethnicity, and disability. The (fictitious) Mimi episodes show diverse characters, including a deaf girl, engaged in the work of navigating the ship and doing the research, and students identify with the different characters. The case study data (Middlebrooks et al. 1996, p. 491) " . . . capture clearly Mimi's success in appealing to students, regardless of their gender, race, ethnicity; geographic location; grade level; or school achievement." The study quotes a teacher as being representative " . . . when she praises Mimi for offering not only a 'wonderful science unit,' but an opportunity to cover the 'differences' that are part of real life." One disparity documented by the case study researchers is that boys tend to use the computer-based materials more than do girls. Moreover, they note that, while the project created episodes demonstrating diversity, the working scientists shown in the actual expeditions illustrate that the scientific enterprise is in fact largely white male. According to the classroom observations conducted for the case study, teachers feel able to extend the Mimi episodes by discussing age, gender, and disability, but not ethnicity, class, or culture. Nevertheless, among the U.S. examples of innovations we studied, Mimi goes as far as any in dealing with issues of diversity.

The Comprehensive Projects

The basic assumption behind UMC was that "well-connected and professionally active teachers would be able to improve upon the poor mathematics performance of inner-city students" (Webb et al. 1996, p. 248)—certainly an equity goal. Nevertheless, the focus of the innovation, and hence of the case study, was largely on the means toward that end—creating conditions for teachers to become active professionals well-versed in the reforms being advocated in mathematics education. The project developed an equity paper, and individual teachers made strong commitments to the proposition in the paper that all students could learn challenging mathematics, though their approaches to attaining that goal differed. The teachers themselves reported successes with their students, but the case study is silent on whether progress was being made on the ultimate goal of raising student achievement in the participating districts. One problem that concerned teachers in some of the collaboratives was the lack of success in reaching minority teachers and teachers in largely minority schools—expressing the same concern at the local level that the Ford Foundation had tried to address at the national level with the UMC project.

NCTM's ultimate goal in developing the *Standards* documents was to get more students to study more mathematics. Issues of equity and student diversity are not specifically addressed in the documents, however, leading some to criticize them on that ground. The case study describes what happened to the efforts of the writers drafting curriculum standards for middle school. They wanted to include some statements about cultural influence on learning: "' . . . what is in the personal culture of each child that is mathematical' in the words of one writer" (Webb et al. 1996, p. 54). But the majority of the NCTM *Standards* writers feared a loss of their main emphasis on content if the whole complex array of needed reforms was addressed, for example, specifics on how to reach underserved populations. Equity made its strongest showing in the mathematics *Standards* in the general call for equal access and opportunity for all students.

The approach of Project 2061 to equity issues is discussed by Atkin et al. (1996b, pp. 228-32) in terms of inclusiveness and diversity. Their case study points out that project materials stress the importance of creating learning goals that, though challenging, make sense for all students. Both *Science for All Americans* (AAAS 1989) and *Benchmarks for Science Literacy* (AAAS 1993), the project's landmark publications, envisage that all students will learn the same content, no matter what their background. One way that the project has addressed inclusiveness and diversity goals is through commissioning a blueprint on equity, to be published in 1997 together with blueprints on other implementation issues. According to the case study, however, the goal of inclusiveness is not discussed in any great detail in the existing documents—what such a goal might imply about direct targeting of minorities and females, for example. As to diversity, the project selected the original six cooperating school districts on the basis of location (rural, suburban, urban) and the ethnic makeup of the school population. This strategy paid off with respect to alternative curricular approaches that, although not necessarily sanctioned by the project's central staff, took root in some of the sites. San Francisco, for instance, has made a special effort to provide science learning experiences and role models that reflect students' diversity and interests, including the infusion into the curriculum of scientific contributions by historically underrepresented groups. The case study concludes the discussion of equity issues in Project 2061 with a question: To what extent will equity be considered a matter for national attention, and to what extent will it be addressed primarily at the local level? In the past, it has been mainly the latter, " . . . but a project that continually emphasizes its commitment to all Americans has created expectations that seem to suggest the necessity for turning significant resources toward the issue in the near future" (Atkin et al. 1996b, p. 232).

California is one of the most diverse states in the nation, if not the most diverse: the state's schools serve over 100 different ethnicities and linguistic groups. It is not surprising, then, that the rhetoric of equity pervades the California science education reform effort. But according to the Atkin et al. case study (1996c, p. 100): "By the description of the reform leaders, however, this rhetoric has not translated into broad successful action." The 1990 *California Science Framework*, the document that can be said to have launched the reform, does contain a section—probably a first in state frameworks—on effective teaching of science for women, minority students, and students with disabilities. Yet the leader of CSIN, the elementary teacher network created to implement the *Framework*, told the case study researchers: "The equity issues were not a priority in the beginning of CSIN . . . we were fighting just to get good science established . . . That was really our focus and target" (Atkin et al. 1996c, p. 101).

When CSIN did confront equity issues, CSIN leaders were surprised by the negative reactions of many of the teachers and teacher leaders. Nevertheless, it would be wrong to give the impression that no efforts are under way to make science education more effective for California's minority students; the case study describes in some detail the activities of one group associated with the California reform, the California Science Project, to integrate equity concerns with improving science education in the state. The case study concludes its discussion of equity issues in the California reform by considering two different approaches: the bureaucratic one of devising models or programs in some given settings and then applying these to many other settings, as contrasted to the particularistic one of programs developed individually within communities, a course urged by some reformers experienced in the field but one not necessarily consonant with the current emphasis on standards-based reform.

Equity Issues and Science Education Reform

A full commitment to equity and diversity includes special treatment for those of different talents and backgrounds who can profit from it, not only to honor them and increase their access to the mainstream but to enrich the societal talent pool. How to make this commitment become a reality in the classroom has been a difficult challenge for science education reformers. In the view of many, tailoring the science curriculum to various cultural and ethnic groups inevitably dilutes it in ways that clearly signal lowered expectations for the students from these groups. Moreover, it is nearly impossible to accomplish such tailoring on a centralized basis for all the different groups represented in the nation's schools, leaving it to the classroom teacher to make the desired adaptations—another near-impossibility. To help teachers in designing science instruction effective with their students, equity issues sometimes are defined in terms of learning styles. Apart from the tendency toward stereotyping, such an interpretation makes it relatively easy to argue that constructivist approaches applied to the class as a whole will take care of the students' differing backgrounds.

One justification for special curricular attention at the local level to gender, class, and ethnicity lies in the fact that school improvement and educational reform depend on good understanding of teaching and learning by students, parents, and educators. No single grand perception is shared by all people, especially across cultural borders, because school problems appear to most people to be "situated": that is, the perceptions of good curriculum and instruction take on the characteristics of the people in a given neighborhood—the social, economic, and political features that appear linked to that locality and time. To an outsider, many problems look the same from school to school; but to an insider, they often appear unique, tied to the special mix of students, teachers, administrators, parents, and community and tied to the special history, culture, politics, economics, and aesthetics of the school. The same curriculum often does not work the same in similar schools. When good teaching depends on a fit with local conditions, it is important to facilitate curricular choice by those best acquainted with local conditions.

The matter is not just curricular. The democratic ethos holds that the human condition is served by vesting power in the people. Nurturing the sophistication of the people to exercise that power is the responsibility of all reform initiatives. It is easier, however, for reform projects to leave education of the public for others. Defining benchmarks of literacy and competence independent of local and personal circumstances is common in the contemporary world of U.S. reform of education. For a decade, the common advocacy has been for "standards," as discussed in the case study of the NCTM *Standards*. Some spokespersons argue that investment in benchmarks and academic standards is an expression of support for those children furthest below the standards, particularly the "special populations." Yet it might be asked whether the current movement is but a variation on the theme of minimum competency and statewide goals running through the states for 20 years (Coley and Goertz 1990). Schools whose students are found least literate have been subject more to embarrassment than to help, have been constrained from deciding how to serve their own communities, and have seldom been given special funding or other provision for higher quality instruction. In spite of the fact that many claiming to speak for equity disdain the idea of situated standards, concentration on literacy without dealing with the special problems of low resources and cultural diversity is weak support for equality of educational opportunity. Given the size of the equity problem, we found its role in the eight innovative programs to be underplayed, as revealed in the case studies of which they were the subject.

Some Examples From Other Countries

Two of the countries participating in the OECD case study project, Ireland and Canada, focused on innovations dealing with gender equity in physics instruction. The two cases form a study in contrasts, one having taken a centralized, bureaucratic approach, the other a highly particularistic one. We summarize both studies below.

Ireland. Ireland has a highly centralized system of curriculum development and assessment; however, subject offerings are at the discretion of the individual school. Many of the senior secondary schools are segregated by sex. In the past, a number of the schools serving girls did not offer physics, chemistry, and advanced mathematics. The lack of provision was particularly severe in the case of physics. As a result, 80 percent of the boys but only 33 percent of the girls were in schools where physics was taught. A decade ago, the department of education introduced a scheme by which experienced visiting physics teachers (later also chemistry teachers) co-taught this subject with the resident science teacher (usually a biology teacher) in recipient schools, until the resident teacher was able to take over. The scheme worked well as a form of professional development and did result in increased enrollment of girls in physics—about 10 percent over an eight-year period. During the same time, enrollment in chemistry decreased for both boys and girls, but much less so for girls, so that the boy-to-girl ratio in this subject decreased from 1.67 to 1.25. (The ratio in physics changed from 3.9 to 3.2.) The case study researchers presume the falling enrollment in chemistry was in part due to the greater availability of physics. After making physics available to girls, their performance on the terminal high school examination matched national norms. Pertinent to equity issues is the observation in the case study that neither the physics curriculum nor classroom instruction changed materially in the transference of the physics course from boys' to girls' classrooms. Even though the intervention resulted in a somewhat higher enrollment of girls in physics and their performance was acceptable, the case study concludes that gender equity requires reconsideration of the physics curriculum and instruction, not simply providing access.

British Columbia, Canada. This case study also concerns gender equity in physics instruction, specifically in an electricity unit in grade 10. The setting, however, is a single classroom, even though the impetus for the intervention came from both the provincial ministry of education and the national level. The classroom was not typical in that the teacher, a woman with a Ph.D. in immunology (only 18 percent of the science teachers in British Columbia are female) was committed to creating gender equity in her science classes. The case study notes that she actively encouraged girls to participate and talked about her own experience as a scientist, woman, and mother. The focus of the intervention was to make explicit gendered ideas about science—"physics is harder for girls than boys," "boys like physics better," "girls will be isolated—seen as 'nerds'—if they take physics," "girls don't really need to know about electricity," etc.—and have girls and boys discuss them openly. In addition, the 12-week electricity unit was redesigned to embed the science content in social contexts, so that after investigation and study students could deal with decisions relevant to the contexts. An assessment at the end of the unit included some items that allowed considerable latitude in response. There were no differences in the test scores between boys and girls, but in the open-ended items relating a scientific topic to a social issue, more girls than boys tended to emphasize social, behavioral, and environmental factors, whereas more boys tended to take a technical approach. The case study recommends that assessments, to be fair, should accommodate these alternative approaches when appropriate. More broadly, the case study found that making gendered ideas about physics visible through discussion in a supportive atmosphere and adapting the curriculum to meet girls' as well as boys' interests proved to be successful strategies in enhancing gender equity. The study also raises some caveats, however: pressure on girls to enroll in physics when they are not really interested, and resistance on the part of the boys to the emphasis on gender equity. Nevertheless, the study concludes that the approach used in this one classroom should be considered for other science classes and developed to suit individual circumstances.

Summary

Probably there are no easy reforms of anything. Educational reform is no exception. The stories in the eight U.S. case studies illustrate the long and costly effort of making even small changes in the teaching and learning of science and mathematics. Although the National Science Foundation and many others have called for systemic reform, making broad, simultaneous and interrelated changes, there is a limit to how many causes the reformer can take up at once. Undertaking grand changes in curriculum and pedagogy, the reformers in the innovations we studied were found to be little engaged in simultaneously upgrading assessment, program evaluation, and equity. We have noted, case study by case study, the endorsement of change in assessment of student achievement—but little participation in it. We have noted little reliance on formal use of program evaluation and the growing expertise on change processes. And we found few efforts to adapt to differences in background and styles of students, teachers, and citizens. These were underplayed issues in the eight case studies discussed in this volume of *Bold Ventures*.

7

Assessing the Implementation of Innovations in Mathematics and Science Education

Michael Huberman

As noted in earlier chapters, the eight innovations studied share some important features. First, they represent a reformulation of the science and mathematics judged worth teaching in school. Within that broad mandate, all put less emphasis on traditional disciplinary knowledge and more on linkages across subjects, on socially relevant topics, and on scientific and mathematical derivations related to real-world or practical problems.

Second, they espouse an altered view of learners, who are made more responsible than previously for the construction of scientific or mathematical meanings in the problems under study, whose working procedures are more conceptually mobile, and who are more interactive in working with the understandings and explanations of others.

Finally, all eight projects construe teachers as having a deep understanding of the processes at work in pupils' learning and of the linkages between pupils' cognition and the content of mathematics and science. A primary task for teachers, then, is to present activities that facilitate—rather than shortcut or overpower—the processes whereby pupils develop their own conceptual scaffolding in science and mathematics. In the Precalculus project, for example, one no longer begins with the concepts and procedures to be taught, then demonstrates how to apply them. Rather, one begins with an applied problem and derives the underlying mathematical concepts and functions.

These core purposes of the eight projects constituting our sample led to highly variable adventures in states, districts, and—above all—schools. From the outset, the projects were different species in terms of their respective settings, sources, and magnitudes. In this chapter, these differences are respected while we attempt to identify genuine intersections and commonalties from which actionable lessons can be derived.

The Macro and Micro Innovations

In this chapter, we examine the comprehensive projects and the curriculum development projects that comprise our case studies. In their design and implementation, the comprehensive projects are depicted here as ambitious, audacious undertakings, affecting different levels of policy and practice—often before their time. They thus enter into a "time warp," where few policies and practices are ready to accommodate their way of construing and implementing science and mathematics education. The issue is to anticipate the time warp, yet push ahead with innovations.

We track this progression in the eight projects, looking first at the adoption process in local settings: key advocates, accelerating factors, and impediments; the roles of program developers and publishers; leadership roles; and perceived uncertainties and concerns on the part of teachers and administrators. We then examine in more detail the issue of teachers' substantive backgrounds in science and mathematics and their belief systems, as well as the implications for the reforms undertaken. Next, we focus on the nature and magnitude of demands made on institutions and actors by the reforms and innovations—that is, the burdens of change. This leads to an analysis of the process whereby local mastery of a new instructional approach may be achieved, more general teaching capacities enhanced, and evidence for student outcomes and longer term institutionalization of innovations measured. The key to such measures is the close study of enactments—the actual implementation, on a daily basis, of the innovations in garden-variety classrooms throughout the country. It is the shortfall between policy and enactment that exposes the actual extent of innovation at its most crucial level—at the actual locus of instruction and learning.

To provide a preliminary view of the differences among the projects, we briefly sketch some defining features of each. There are, first, the comprehensive innovations or "macro ventures." They can respond to a more ambitious strategy for large-scale change by:

- operating at a more macro and/or systemic level;

- granting more importance to the state;

- aligning curriculum and accountability;

- focusing on some privileged areas (technology, equity, staff development); and

- forging tighter alliances among professional associations, universities, and key citizens' groups, including business and industry.

In some instances—for example, the California science education reform—it would appear that all of these were happening simultaneously. The California reform undertook initiatives at virtually every level from professional associations to classrooms and created specialized roles for affiliated teachers and training for administrators. This strategy of multiple interventions at multiple levels—this sort of "particle cloud" concept of reform—is a particularly promising one.

In the same macro sample, we have Project 2061, a long-term redefinition of science education policy and instruction, with six loosely affiliated experimental sites and a set of achievement indicators or "benchmarks" at successive grade levels. A third reform effort, the National Council of Teachers of Mathematics (NCTM) standards project, moved deliberately from a policy forum to publications that progressively approach local instruction and assessment while accommodating a wide margin of local interpretation and adaptability.

Midway in scale, the Urban Mathematics Collaborative (UMC) project provided technical expertise, interpersonal support, and reflective stimulation to secondary level teachers across the country, notably in more disadvantaged areas. Some 15 years after their establishment, all of the collaboratives were operational but at varying levels of participation and financial stability. From all indications, participants were like-minded teachers. The number of epiphanies or changed visions of practice reported in the case study suggests actual changes in capacity and practice. It is unusual to find such a "marginal" project (i.e., one located outside the school system) with such widespread and significant impacts at the instructional level after a fairly short, but intensive, involvement by participants. This calls attention to the nature of the strategy put in place at the collaboratives.

We then have the more micro-level innovations to account for. Three of these—the Voyage of the Mimi, ChemCom, and Kids Network—are imports from professional associations, development centers, and publishing houses; but did involve many teachers in their genesis, field trials, and instructional formatting. On the other hand, Contemporary Precalculus is a genuinely home-grown product of the North Carolina School of Science and Mathematics (NCSSM), a public coeducational, residential high school for juniors and seniors with a strong potential in mathematics and science. From a dynamic development project, it quickly became an "export" through the publication of a textbook and a series of training events.

Tracing Enactments: First Steps

Comprehensive Projects: The Audacity of Reform

It is possible to compare some of the macro innovations in tandem with the micro innovations. Many projects, in fact, combine both macro and micro features. For example, Project 2061 had six experimental sites throughout the country. The NCTM *Standards* and successive documents have, in a short time, become an anchoring point in hundreds of school districts—one source claimed a 40 percent readership for the *Curriculum and Evaluation Standards*. The California reform was already a reference point in dozens of schools with the aid of staff developers, consultants, lead teachers, and administrators-in-training—many still active in their classrooms. In other words, the gaps between policy and ongoing instructional practice appear to have been reduced considerably in time and space.

This fact is, in and of itself, significant. In effect, the conventional time frame for such audacious or ambitious ventures is millennial. The original work by Paul Mort and his associates at Columbia showed a 50-year gap between the initial use of an educational innovation and an informed awareness by 30 to 40 percent of the appropriate teaching staff in the proximate region (Havelock 1969). Moreover, reformers are not interested in just a handful of districts or experimental projects, but in getting to critical mass.

This foreshortening of the time frame has advantages and limits. For example, to reach critical mass, several distinct "planets" (state and local administrators, local staff developers, professional associations, university teaching and research staff, publishers, mathematics and science educators, business representatives, citizens' groups, and—of course—local teachers and pupils)—each of which tends to orbit on its own axis—must be coalesced. In the reform cases studied here, they are orbiting one another in real time, sometimes for the first time, with a progressively more common discourse and agenda.

Anticipating later discussion, we would conjecture that, in this situation, the California logic may be a promising one: the "particle cloud" approach to reform, with a swarm of small to mid-size initiatives inhabiting and linking each of the constituent groups. Thus, workshops, courses, or summer institutes are backed up with means for networking and other informal contacts, making good use of the technologies at hand. State-level policy is then improvised in light of anticipated problems, while still attending to interpersonal links and antipathies and gradually creating a new Zeitgeist for science. The particle cloud strategy might make the average teacher believe that *this* reform is not going away so quickly as it pervades, invades, territories on all sides. Soon, Project 2061 will have our teacher studying its *Benchmarks*; NCTM will be influencing the teacher's district's assessments. The *California Science Framework*, homing in on appropriate grade ranges for its inquiry-based approaches, has our teacher's grade in its telescope. The teacher's administrators will have become substantively knowledgeable through the California Science Implementation Network (CSIN), and the mix of trainers and consultants—many of them former colleagues of our teacher—will have lent credibility and legitimacy to the enterprise. So superficially, at least, this constellation does have the means of setting off measurable effects of state policies on the classroom, but not with a top-down scenario. Rather, actors at all levels are surrounded by a welter of convergent, but only partly engineered, initiatives.

Curriculum Materials Projects: An Apparent Loss of Audacity

We may have overstated the rapidity of diffusion of the reform projects, although the case study data support this interpretation. For the micro-level innovations, however, the data suggest a generally countervailing trend.

For example, in the cases of Voyage of the Mimi and Kids Network, developers and/or publishers appeared to deliberately attenuate some of the instructionally most challenging strategies for pupil experimentation and "meaning making." Mimi combines videos, computer software, and print software around an integrated set of concepts in mathematics, science, social sciences, and language arts for the middle school grades. It is well-conceived, the product of a lengthy, expensive development process. It addresses real-world issues alongside scientific content. For example, the first voyage incorporates the study of proportional reasoning, triangulation, and navigation, while integrating issues of diversity and equity. The second voyage consists of an archeological expedition in the Yucatan Peninsula of Mexico, taking up the Mayan number system, the Mayan calendar, the relationship between the Earth and the sun, and a discussion of social issues arising during the trip. Mimi is a multimedia package (television, print, and computer materials), primarily focused on science and with an activity base that promotes hands-on exploration and independent exploration.

And yet, after field tests and pilot training, Mimi came to rely mainly on classroom print materials—pupil activity guides and scripted teachers' "overview guides"—both constituting a lesser reach for most teachers than the less prescribed hands-on experiments and computer simulations. The overview guides came with questions for specific episodes, along with complementary information and suggested activities (e.g., instructions for building miniature lighthouses). There were, to be sure, some extending concepts in Mimi, interactions through which pupils could try out ideas, hear the ideas of others, and evaluate their own reasoning. Still, there was a discernible dilution in the scope and risks of the project. As Sam Gibbon, the program developer, noted, "We wanted teachers to welcome opportunities to say, 'I don't know.' [We] underestimated how hard it is for teachers to feel out of control of the intellectual sequence, the 'What are we going to learn next?' question" (Middlebrooks et al. 1996, p. 419).

According to the case study, most teachers stuck closely to the prescribed guides, using instructional and management skills they had carried with them through the years. But it is worth noting that some had never worked before with "expeditions" or computer simulations, nor with the replication of scientific experiments. This, in itself, was an instructional stretch. Also, a few teachers wandered gradually from the guides, inventing their own questions and experiments.

Regarding ChemCom, we find a similar paradox. The developers repeatedly emphasized the importance of active participation in decisionmaking, data analysis, and group interactions, but the "Comprehensive Guide" accompanying the curriculum is laden with guidelines; tips; expected results for laboratory activities; and advice on instructional management, activity formats, and tests. Once again, it would appear that most teachers were deemed not quite ready for deeper and wider changes in mathematics and science, as represented by the reform projects we have just reviewed. Both Mimi and ChemCom may thus be seen as "transitional" projects that incorporate new priorities but are designed or redesigned for a more timid initiation into the classroom. Unlike the newer, bolder reforms, they have been relegated to a slower lane.

Kids Network presents a variation on the same scenario. The developers had in mind a set of experiments dealing with important social and scientific themes (Trash, Weather in Action, What's in Our Water?, Solar Energy, etc.). Pupils in grades 4-6 were to conduct cooperative experiments on each theme, then use a telecommunications network to send their local findings to a central computer which pooled and analyzed their data. The analyses were to contribute to new questions and interrogations with peers and with practicing scientists via the network. Again, there were both a detailed guidebook for teachers and pupils and an attractive, varied set of accompanying materials (videotapes, a software manual, lab materials with equipment, maps, and activity sheets). Gradually, formats prescribed in these materials became the norm.

Consider the Kids Network unit Acid Rain. First, students recorded the names of different foods at home and the acids they contained. They then followed directions for measuring the pH of different liquids, designed and built a rain collector, and explained the differences in the pH readings they recorded. Afterward, they examined a map of acid-producing gases in their area and wrote a letter on the topic to their telecommunications "teammates." Other activities included the observation of objects immersed in liquids at different pH levels, the collation of pH readings of rainfall into a chart spanning three weeks, the simulation of settings likely to have the most and least acidic rainwater, and the prediction of pH levels of sample objects (shell, paper clip). In terms of content and instructional potential, this was—again—a promising project.

Initial evaluation data and field testing for Kids Network were equivocal. Teachers tended to dominate discussions that synthesized findings. Conducting the investigation was emphasized more heavily than interpreting the results. Many students perceived the activities as overly conventional, even the hands-on inquiries; some wanted to try out their own ideas or communicate more with other schools. For the developers, the balance of teacher-led and student-led activity was elusive—but it swung toward a prescriptive, teacher-led curriculum. Still, for many teachers, this was the first time they had the opportunity to experiment with investigative curricula, with the elaboration of children's understandings through group interactions, and with thematic or socially relevant topics. Some experiments allowed for alternative procedures and unknown or alternative solutions. This component, along with the telecommunications, constituted the "stretch" in many teachers' representations and repertoires. For the most part, however, the case study researchers found the codified procedures to be carried out in uniform ways by students at the six sites they observed.

At the outset, then, Kids Network, ChemCom, and Mimi looked adventurous in terms of being carefully wrought, imaginative scientific and mathematical development projects. Once through the production cycle and into the mainstream classroom, however, they appear to have lost their bite—more so, in fact, than the macro reforms. To extrapolate, cautiously: Stand-alone innovations may be less insulated against dilutions or trivializing or downright reversion to conventional practice than reforms whose local enactments are hitched to new policies and new organizational arrangements at district, state, or national levels. In other words, "systems" thinking does appear to make a difference in preserving a steeper gradient of change—providing that this thinking is not hyper-rational, but rather accommodates the unknown, unpredictable, and Murphy-lawlike nature of institutional life. The particle cloud strategy noted with respect to California seems well-equipped to deal with the vicissitudes of the contexts within which educational policies and practices exist.

One innovation, however, breaks these rules. In part, this is attributable to the exceptional conditions enjoyed by the Precalculus course at NCSSM. Here, the case study team found a self-selected, highly motivated, well-trained faculty; a resolute and dynamic leadership; a cohort of students hand-picked for their interest and abilities in mathematics; norms of mutual support and collegiality that encouraged risk taking among staff; an open syllabus; a brief to challenge colleagues' thinking during program development; and, finally, a mandate to export their product as they saw fit into other environments. Add to this an amenable publisher, and it is easy to understand why the Precalculus project was more able to preserve its problem-solving approach to the derivation of mathematical models and functions, without backsliding into the memorizing of formulae or the practice of purely computational skills. This appears to be the story of a constructivist curriculum that, in the course of its development and publication, made few or no compromises.

Once outside NCSSM, however, the "rules" we hypothesize begin to apply once more. In effect, all did not go smoothly at the five schools observed by the case study researchers in which the NCSSM precalculus textbook was used. Still, there was less haggling with the publisher than is common, and fewer frictions than usual within the—somewhat atypical—adopting schools.[1] Also, the financial and institutional pressure to diffuse the Precalculus course was less than is the case for many reform projects. It might even be claimed that the prime objective for adoption at other schools was as much to feed back corrective information to the developers at NCSSM for further experimentation as to spread the textbook.

This is a different context than trying to introduce the Precalculus course to the mathematics departments of teachers attending the Urban Mathematics Collaborative, or to the upper level mathematics teachers in the California

---

[1]The schools generally were in the forefront of various reforms; for example, the investigative pedagogies were already familiar to them.

school districts subscribing to the *Science Framework*, or to the 9th and 10th grade teachers in the schools described in the NCTM case study. For now, we note that scale (macro/micro) may not be as decisive as the contexts that are connected in policy and execution—often through coalitions or via associations—such that organizational and instructional risks are actively solicited, taken, and maintained, and where program developers and teachers have sufficient freedom and appropriate supports to maintain these risks through to local consolidation of the innovations.

Reforming in Time Warps

As designs diffuse regionally and locally, then, they confront new decision points, many leading to successive dilutions or conservative modification in science and mathematics education. Note, too, in passing, that the advocacy of local teachers and administrators is crucial, but cannot be taken for granted—far from it, in fact. Not all actors involved in the cases under study were advocates of the reforms, and many had good reasons not to be: long-term uncertainties; possible or actual loss of status, influence, or power; redefinition of their posts. For many, there were logical grounds to resist, dilute, or defend themselves against unwelcome changes. Moreover, since many of the Project 2061 benchmarks and NCTM standards were sufficiently broad, the specific links to learning activities were largely left to individual schools and teachers. This made dilutions easier and more legitimate. The reform drift, however, was clear—although it had not yet gone mainstream in the settings under study here. Where it was stronger and more constraining, as in some phases of the California reform, there were more direct pressures on actors at all levels to come deeper into the new paradigm.

According to empirical studies of the process of change, the strategies of "traditionalists" are not typically passive ones. It matters less that they are opposed, say, to integrated science or investigative mathematics than that they are actively committed to other perspectives. They do not always hide under rocks. Nor are they resistors, but rather protagonists of another way of seeing or doing things. Moreover, they have learned well how to use formal structures and informal working arrangements as buffers. In our eight case studies, we mainly observed actors who have espoused a set of gradually more influential trends, the big ideas in contemporary mathematics and science, and who are seeking to root them in the classroom. We need to remember that there are countervailing structures and people who think and act otherwise, and that they were—and probably still are—far more numerous and equally powerful.

This makes for a "time warp" phenomenon. For example, developers in mathematics and science education appeared open to the new perspectives laid out in earlier chapters. Still, they tried to anticipate the extent to which teachers would tolerate the new approaches or their instructional demands. Publishers could be equally sensitive to new paradigms, but they were usually after the largest market. At the same time, the nature of core concepts in mathematics and science education was—and still is—in constant transformation, and there continues to be a mix of perspectives in local practices. Note that one paradigm, one set of integrated perspectives, is not necessarily "better" than another, but it does claim to address or resolve issues better than its predecessor. The projects in this study were variations of an emerging paradigm for mathematics and science education: new ideas that have gradually commanded more of educators' attention and commitment. Gradually, too, they have emerged as dominant perspectives—dominant, but always at war with their predecessors and successors. Also, and to be more precise, many of the new ideas in the emergent paradigm in mathematics and science education were present some 30 or 40 years ago, but were then considered marginal, impractical, or premature as possibilities for change on a national scale.

Consider Kids Network. Over the past decade, it has subscribed largely to a "constructivist" perspective among researchers and teacher educators, but to a more cautious, "experimentalist" or hands-on approach among program developers. In prevailing classroom practice, the paradigm has been more conservatively didactic, as depicted also in Voyage of the Mimi. In fact, all three perspectives are represented in the eight case studies, within and between policy levels and practice settings. Thus we find the inevitable time warps in the ripening of these different scenarios in different settings and at different levels—with the most difficult level to change at the actual locus of instruction and learning.

At present, we can discern a widespread, experimentalist version of Kids Network as reported by the case study researchers. Gradually, this version replaced science lessons that consisted of watching a film, reading a textbook, filling out worksheets, writing a report, or listening to a lecture by the teacher. In well-equipped classrooms, teachers carried out demonstrations or even replications of classic experiments. At the end of a topic, students did not always take paper-and-pencil tests to assess their understanding of the information. It may well be that the modal practices in middle school science and mathematics are not yet at this point. The Kids Network sites, however, are close to this point of generalized instructional practice.

In the succeeding instructional scenario, illustrated by some classrooms studied in the Voyage of the Mimi, students routinely did some hands-on, rather than purely vicarious, investigations requiring them to physically manipulate and account for the properties of everyday objects. Generally, however, these investigations were algorithmic in nature. Teachers provided materials, directions on their use, and a system for recording observations. While working in small groups, students typically followed the same set of directions for experimenting with the preselected materials. "Correct" investigations led to the same results among groups. Teachers predominated, both procedurally and analytically. Preset lines of inquiry, with faithfulness to directions and uniformity of results, meant that there was greater emphasis on the conduct of the investigation itself than on the analysis and discussion of its results and their significance.

Once again, we find a paradox of perspectives. In the conduct of instruction, Mimi classes we observed were nontrivial departures for the teachers involved. If one looks at Mimi's actual design, however, many aspects *have* been trivialized. The overall picture is still more complex. Even though the scientific investigations were prestructured, from a teacher's or pupil's perspective, they could be adventurous. They sometimes contained unknowns and obstacles: interpreting the directions accurately, setting up the materials correctly, recognizing group or individual errors or misconceptions, reconciling procedural differences, knowing when answers were plausible, especially in light of challenges from classmates ("that's not the way *I* did that test"). Furthermore, students often generated new ideas from participation in the more standard investigations.

What is observed here, then, is a transitional paradigm, but one still at some distance from the vision of Project 2061, the interactive inquiries of UMC, the social constructivism of the *California Science Framework*, the investigative pedagogies implicit in the NCTM *Standards*, and—perhaps, above all—the applicative procedures of the Precalculus project. Some of the reforms are conceptually more audacious; some are more organizationally radical. In a few cases, such as the Precalculus project, concepts, materials, *and* the actual settings for learning and instruction have been transformed in apparently durable ways—but only in a single, pilot institution.

Understandably, the most critical changes need to happen at the classroom level. The rest, in many ways, is talk and paper. All eight reforms and innovations entail a close monitoring of students' reasoning along with an exploitation of those often elusive "teachable moments." There are also the connections to scientific or mathematical content matter to establish and an assessment of these links in ongoing activities. For this, teachers must master their mathematics or science at a level of fairly deep understanding.

If, as we have claimed, each of the instructional scenarios observed in the case studies derives from a roughly common paradigm, there is still a wide continuum. To further complicate matters, this continuum may hold for different classrooms and certainly within and across districts. To look closely at the science and mathematics instruction within a given classroom, in fact, one may see side-by-side practices from traditional and constructivist perspectives (Cohen 1990). Paradoxically, this reinforced some teachers in the belief that they could continue to teach conventional mathematics and science, since their "progressivist" colleagues were doing it as well.

In this respect, the cases examined here act as an accelerator. For example, the Kids Network program was eventually designed for a generation of science teachers who could execute the units in a more algorithmic, structured hands-on curriculum. This represented a serious watering down of the initial design. The most likely users of the program, it was thought, were unready for more radical, constructivist designs. As teachers worked with the program, however, some practices became less embedded in this model. Many teachers then found themselves intrigued with revisions that entailed more actual experimentation. For lack of external tutelage and support, however, their actual efforts were far more modest than their critiques of the program.

We have claimed that the more classic scenarios for science and mathematics education were generally the dominant ones in absolute numbers, even in the innovations studied here. The more comprehensive reforms bear witness, however, to a more resolute policy shift. At the same time, instructional practices seemed to be moving at a glacial pace. A critical feature of this landscape is the gap between points on the continuum from "conventional" to "constructivist" instructional practices in the different scenarios depicted above. If the next point on the continuum appeared unreachable to potential users, we witnessed, at best, incremental change. As exemplars of a sort of dynamic conglomeration along the continuum, the Kids Network, Mimi, and ChemCom are thus critical barometers for researchers and practitioners alike.

Adoption of the Innovations

Scholars of the adoption process (e.g., Rogers 1983) have long noted that the term includes smaller steps which, for any potential user, can stop at any point. A district coordinator, for example, could hear that ChemCom is apparently an interesting project—then forget about it. Or she could go farther and inform herself. If she likes the looks of the project, she could convene her colleagues, and weigh its merits and demerits, then perhaps conduct a trial run of one of its modules. Assuming a positive outcome, there might be an engagement to implement the project.

But "adopting" the California *Framework* or Project 2061 or even the NCTM *Standards* is a more slippery proposition, since it is not always clear what specific engagements are implied. By contrast, "adopting" an urban mathematics collaborative implies active participation; "adopting" the Voyage of the Mimi, Chem-Com, Precalculus, or Kids Network is also more straightforward. Presumably, for the larger scale reform projects, adoption implies that a school district will contribute time, resources, and some form of active consideration to the project. In the best cases, there will be an actual application to ongoing planning and instructional practices. For example, the Kids Network was incorporated in the California reform. So the micro innovation became one of the concrete components of the macro reform. More generally, the major reforms have tried to promote specific initiatives that would move a school or a district farther along the continuum of paradigms we described earlier.

Strategies and Vehicles: The Comprehensive Projects

What means were at the disposal of the large-scale reform projects for inducing adoption at the local level? Typically, policy instruments take the form of requirements; they prescribe desired practices through legislation, rewards, sanctions, persuasion, expertise, and advocacy (Porter et al. 1988). For the reform projects in our study, the strategies of adoption (in the sense of getting actual commitment) appeared to be "soft"—indirect and nonprescriptive, probably with the idea of a gradual, less painful shift in local practices—but with an eye to a more muscular follow-through if there was local foot-dragging or distortion. Clearly, there are disadvantages to this strategy, including local misinterpretations, procrastination, trivial implementation, and the legitimizing of poor instruction; but it appears to have had the great advantage of mobilizing few powerful adversaries in the cases studied. There are intimations in the reform projects, though, of more blood-letting earlier in the process, precisely at the academic and policy levels, when standards and frameworks were first being hammered out.

Table 7-1 enumerates the type and overlap of strategies used to further local adoption on the part of three of the reform projects: the California science education reform, the NCTM standards, and Project 2061.

This catalog does not include some of the less obvious variables affecting local adoption: the critical role of leadership (notably in the California reform and Project 2061), the exercise of pressure, privileged access to media, the necessary compromises that affect levels of quality, issues of resource allocation, or the concern that many teachers may not be adequate to the demands posed by the reforms—just to name a few.

The catalog does, however, provide a bird's-eye view of the adoption process at the local level and suggests that disparate reforms shared common strategies of flexibility, some equilibrium between top-down and bottom-up strategies, and—above all—demonstrations that they could fill the middle ground between generous visions and daily practice in a new way. Structurally, this set corresponds to what has come to be known as the "second wave" of school reform in the United States, as discussed in the *A Nation Prepared: Teachers for the 21st Century* (Carnegie Forum on Education and the Economy 1986).

The Adoption Process at Local Levels

How did all this play out locally? First, our data show little or no articulation between the large-scale reforms (Project 2061, California reform, NCTM standards, and—to a lesser extent—UMC) and the smaller scale innovations (Voyage of the Mimi, Kids Network, ChemCom, and Precalculus). When local actors mentioned the reforms (a rare event), it was typically to legitimate what they were trying to do in a specific project. While administrators were more knowledgeable about the reforms, they, too, were caught up in the local projects loosely linked to a larger reform, and with other agendas. Each innovation was mostly self-referential; the links were with other Mimi people, Kids Network people, ChemCom people, etc.

All the same, these projects were sensitive to the kinds of adoption strategies we have just reviewed. In this section, we take a closer look at these. We end by considering our "outlier" case, the Precalculus project. First, we take up

Table 7-1.   Adoption Strategies for Three Comprehensive Projects

Strategy	Projects
Accepting, even espousing, the changing nature of the project over time, including variable time lines	California reform, NCTM standards, Project 2061
Generating visions ("big ideas" . . . "strands") with no specific prescriptions for curricular and instructional practice; promoting local adaptiveness	California reform, NCTM standards, Project 2061*
Creating "mid-level" projects (Project 2061 and its experimental districts; California's teacher networks: CSIN and Scope, Sequence, and Coordination) and new documents to demonstrate that the reforms can accommodate an ecumenical definition of the new curricula, instructional formats, and learning opportunities, thereby achieving local consensus. For example, the Project 2061 *Benchmarks* are operational at an intermediate level between conceptualization and instruction. They make knowledge statements that are clear, and they approximate the language students themselves could use. They are interconnected; they demonstrate their research base. They are not exhaustive, and they show why a given benchmark has been selected.	California reform, NCTM standards, Project 2061 .
Involving teachers in the design, authorship, or critique of key documents or program components; thereby promoting local ownership	California reform, NCTM standards, Project 2061
Involving teachers in the promised infrastructure for preparation, ongoing assistance, and site management	California reform
Creating new coalitions (state authorities, professional associations, research and training institutions, public and private enterprises, key citizens' committees) to widen the base of adhesion	California reform, NCTM standards, Project 2061
Exploiting the multiple networks by which key information passes, key agreements are made, locals are connected to more "cosmopolitan" sources of advice and advocacy, and trust is created—the underlying trust being that local professionals can implement a reform in areas in which many have never ventured	California reform (especially), NCTM standards, Project 2061

*Note, however, that NCTM and Project 2061 have variably specific standards, which are developmentally organized. Note also that the succeeding NCTM *Standards* documents address teaching and assessment standards, which are far closer to the classroom. Generally, however, the *Standards* are detailed enough for guidance, yet supple enough for alternative pedagogical paths.

the roles of developers and publishers. Next, we look at the actions of advocates and "accelerators." We then turn to the various local incentives for adoption and contrast them, finally, with uncertainties at the field sites in Mimi, Kids Network, and ChemCom.

Developers and Publishers. The most enterprising—and perhaps instructive—example illustrating the role of developers is ChemCom. The project has a long-term sponsor, the American Chemical Society (ACS), whose leadership has helped with design, production, initial diffusion, ongoing resources, and wider dissemination. Few innovation projects have such supports, or have them for as long or at such levels of funding. And few sponsors maintain that level of support in the face of enduring critiques.

Printed materials and newsletters from ACS, "drive-in" awareness workshops run by ChemCom teachers, computer bulletin boards, special booths or programs at professional meetings for educators, and summer orientation centers by the publisher made it hard for a high school chemistry teacher not to become aware of the ChemCom project. An interested district could send a group to a sponsored institute run by experienced ChemCom teachers. This is a good example of blanketing the target public. For the Voyage of the Mimi and Kids Network—and notwithstanding the resources of the latter's publisher, the National Geographic Society—these efforts were far more modest. Typically, a Mimi "fest" or other professional meetings and conventions brought the project to the awareness of a lone administrator or teacher, who then brought it home.

Similarly, the actual adoption process in ChemCom often included a pilot year for the initial two or three enthusiasts in a district, in addition to summer workshops sanctioned by ACS, then more targeted workshops for pilot teachers—generally on released time. Workshops provided not only the rationale for the project, or instructional approaches, but also the comprehensive teachers' guide (as in Mimi and Kids Network), with its tips, guidelines, suggested instructional formats, answers to typical questions, and tests for each unit. It is noteworthy that, while adoption was linked to initial preparation in ChemCom, initial users in Mimi and Kids Network were virtually orphaned once publishers had delivered their materials.

Advocates and Accelerators. In many projects, these are often the same species: energetic administrators, enterprising teachers, specialists in the application itself, people committed to the new paradigm. What is astonishing, as much in our cases as in the general literature, is how few advocates and accelerators are required to put an innovation into play locally. In Mimi, school principals often were key advocates. As such, they had the authority to follow through quickly on their initiatives. The school director in the Precalculus project played the same role in another form. In Kids Network, with its strong telecommunications component, the science or technology specialists were the prime accelerators. Table 7-2 presents a resumé of the process at the six Mimi sites.

Note that the "event chains" vary in length from one to four links, and that in three sites (Sites 1, 4, and 5), it was the teachers who initiated the chain, but required their principal's support. Throughout, in fact, the role of the principals was crucial in providing: funding for the basic package and associated hardware

Table 7-2.   Key Actors and Events Leading to Mimi's Adoption

Site	Actors and Events
1	Teacher ➤ District ➤ Teachers
	Teacher pilots Mimi; her enthusiasm fuels local adoption.
2	District ➤ Principal ➤ Teachers ➤ Teachers
	District writes grant and persuades principal that integrating science, mathematics, and technology will bring his school a success not known before; principal reluctantly releases best teachers from classroom responsibilities as mentors to those remaining in classroom.
3	Principal ➤ Teachers
	Principal returns from conference having purchased Mimi; he provides money for curriculum development; first generation teachers are enthusiastic, second generation users receive with mixed feelings.
4	Developers ➤ Teachers
	Teachers attend training with Mimi developers; excited, they use Mimi as way to integrate subject matter areas.
5	(Principal) Teachers ➤ Teachers
	Principal returns from conference with reports of Mimi; two teachers raise funds for training and materials; response of next generation of users is mixed.
6	Principal ➤ Teachers
	Principal introduces to teachers; provides counsel and funding.

(Sites 1, 4, and 5); paid time for curriculum development over the summer (Site 3); scheduling changes (Site 2); counsel for new users (Site 6).

For the principals, the new practice was a good organizational fit, both programatically and in its minimal disturbance to existing working arrangements. They, too, apparently were interested in moving farther along the continuum of innovative practice in science education, as were their teachers. To get a more grounded sense of the process, we provide a few anecdotes from Kids Network (adapted from Karlan et al. 1996, pp. 305-09).

Adoption at Site 1: Urban, New Hampshire. HA (a third grade teacher) thought she first learned about Kids Network through an article in a teaching magazine or a flyer. Nevertheless, her principal introduced it to her and offered his pedagogical and financial support. The first Kids Network user, HA then teamed up with a colleague. Although overwhelmed by her first experiences, HA "loved it," and was "energized by it."

MD was introduced to the program the day he interviewed for a teaching position at the school. When the principal took MD to visit HA's classroom, she was having problems completing a Kids Network telecommunication. That's when MD first played with the software program. Soon after MD was hired as a combined fifth/sixth grade teacher, the principal asked him to examine the Kids Network materials and see if he'd like to use them. MD first attended a Kids Network presentation at a New Hampshire Science Teaching Convention, then ordered the program.

Adoption at Site 3: Suburban, New Hampshire. Before KL had heard of the Kids Network, she participated in a statewide initiative that put Apple IIe's into teachers' homes. She recalls, "I went through the workshop . . . got the computer to my home, and I didn't know how to turn it on." She then increased her expertise with computers. This laid the groundwork for Kids Network when she first heard about it from a fellow church member, the uncle of the Kids Network project manager at the National Geographic Society.

> He dropped a little flyer in my mailbox, and said maybe I could start doing Kids Network with his favorite niece . . . We had to get the money to buy the software. A parent volunteered to buy that for me. Then the same parent said she'd buy the modem. Then we had to get the dedicated phoneline in . . . [O]ne of our parents who works for AT&T said he would come in and run a line . . . [W]ith the principal's support, we managed to get it into the school . . . [with] about four different teachers using it.

Adoption at Site 6: Rural, Vermont. The principal, DF, was introduced to the Kids Network at a technology fair. The developer, Technical Education Research Centers (TERC), was seeking field test sites. According to the principal, he and some teachers "pushed to become a pilot school . . . National Geographic provided the software, a IIgs, a bunch of things like that." A third grade teacher who attended the technology fair with the principal piloted Hello! When What's in Our Water? became available, "the sixth grade teacher played with that." According to DF,

> Vermont schools were just getting into water quality, lead in the water and so on. So the timing of that was actually pretty damn nice . . . the kids used to work with the custodian taking water samples and checking out all those kinds of issues . . . Acid Rain, when that first came out, we actually had some kids do some research . . . did some sampling on Mount Snow and took a look in the area. So that kind of fit in.

Note that principals again played pivotal roles, either in introducing the program to individual teachers, and/or in providing the pedagogical and financial support for other interested staff. Technology specialists, science specialists, and engaged parents were also catalysts. In all cases except one, teachers were free to choose whether to do Kids Network. This is characteristic of the innovations reviewed in this section.

Relative to the comprehensive reform projects discussed earlier, then, we would note similarities with the innovations in the following respects:

- the desire to "come aboard" programatically (constructivist science and mathematics, integrated curricula, social relevance, scientific literacy for all, special sensitivity to issues of diversity and equity) and to devote exceptional resources to that objective;

- nonspecification of the actual practices to be implemented, and the general acceptance of local variability within classrooms or schools;

- nonconstraining nature of recruitment, yet visible support given to those that sign on, especially for technology-driven projects; and

- the involvement of teachers in the design or training components of the project—a seemingly deliberate tactic of shared ownership.

Incentives and Uncertainties. The desire to adopt new curricula and instructional practices is an equivocal one. At the level of the Kids Network teachers, for example, the social relevance of the topics was appealing, yet the proposed activities in the teacher guidelines seemed inauthentic to some. Those lured by the interdisciplinary activities wondered how they would jibe with the program's time constraints (six weeks). Teachers intimidated by the telecommunications component still were grateful to be experimenting with it.

Incentives. Table 7-3 depicts the primary incentives for adoption reported by teachers in the case study sites for Voyage of the Mimi and the Kids Network. Most, if not all, of these incentives overlap with the priorities of the larger reform projects. They are well-embedded in the emerging paradigm, with teachers suggesting that they were ready to take into the classroom some of the same priorities that were hammered out at higher levels of policy. *Whether* and *how* they are actually enacted in the classroom thus become the salient questions.

Uncertainties. Table 7-4 presents initial hesitations toward Kids Network and Mimi. A few Mimi teachers were critical of the scientific components of the program, noting inadequate coverage of essential skills, the likely need for supplementary materials, and mathematics that meshes poorly with the science material in the curriculum. Note, however, that perceptions of inadequate content coverage often hide other factors: a retreat to—or renewed defense of—textbook-driven or drill-and-practice activities; inadequate training and insufficient ongoing assistance; fears of technology-rich projects; and the like. These also are specters in the reform projects. At the classroom level, where proposed changes must come to life, these factors often are present even within programs generally appreciated by teachers.

Table 7-3.   Incentives for the Adoption of Mimi and Kids Network

Incentive	Discussion
Intrinsic qualities of the program	*Mimi:* • "Scientific," serious, high-quality character • Cultural wealth (opportunities to compare cultures) • Personal relevance ("connections to the kids' lives") *Kids Network:* • Relevance, credibility relative to environmental issues • Creation, through telecommunications, of "global" classrooms
Program format	• Interdisciplinary, integrative nature • Flexibility, open-endedness, provision of a springboard for other activities • Appeal of their technological components (television, computing, telecommunications, graphing, mapping) • Good suggestions for activities • Quality and teacher-friendliness of the background materials • Many opportunities for cooperative learning
High appeal to students in initial trials	"Just enough scientific information to interest the kids but not boggle them . . . it says to kids, 'You, too, can be a scientist'" (Mimi teacher)
Good "fit" for teachers, for curriculum, and for present organizational structures	• Congenial "personal fit" with teaching style • Good curriculum fit with units that intersect other segments of the mathematics or science curriculum • In the case of Kids Network, presence of an updated, more inquiry-centered program ("the kids used to groan when I told them to open the science textbook") • Good structural fit that implies organizational adaptability; for example, both programs can work well in self-contained class rooms—for many teachers, this was a vital feature

Table 7-4.   Concerns About Adopting Kids Network and Voyage of the Mimi

• The novelty of the technology raises uncertainties—e.g., new equipment, time needed to learn its use.
• The programs' deadlines present time constraints that compromise the quality of implementation, e.g., data collection, communications deadlines—"insufficient time to experiment with it . . . it looked overwhelming" (Kids Network teacher).
• Telecommunications software is unsophisticated or outdated, making it cumbersome and inflexible (Kids Network).
• Teachers have insufficient computer experience, thereby limiting student access to and interaction with the materials.
• Some teachers regard several of the science experiences as cookbookish or inauthentic—"More like drill" (Mimi teacher).
• Teachers have inadequate background for inquiry-centered science—". . . it takes a special breed" (Mimi teacher).

One final factor is the hesitation of many students to launch into investigative mathematics and inquiry-centered science. (As one of them said: "Just give me a book and the questions.") This was also a feature of the Precalculus project and suggests that those experiences farther along the continuum will only work for many students, especially older ones, when the teachers who change the rules play by them unequivocally, and so long as the preoccupation with content coverage does not predominate.

Leadership

"Leadership" is a slippery term. Moreover, it has shifted meanings in the new paradigm of constructivist science and mathematics education. It has become a more collective concept, in the sense of team direction, collective regulation of school policy and management, and a view of leaders as emblematic of their constituents rather than as their shepherds. In keeping with "third-wave reforms," the more comprehensive projects were eager to share leadership in order to build larger coalitions reaching around and down into the classroom level, and the administrators in most of the smaller scale projects were ready to cede terrain relative to the demands made by innovators on substantive—and sometimes purely political—issues.

Nevertheless, leadership is a meaningful theme across the eight projects. In some cases, it shifted over time (NCTM standards, Kids Network, California reform). In some projects, the person in the leadership role was a defining factor; in others, a dominant agent was not evident. And there was, as noted, a trend to decentralize substantive leadership to teachers playing a variety of roles, with administrators looking after the more organizational and financial aspects. This was especially the case in the smaller scale projects, as reflected in table 7-5.

Our examples constitute a varied set, especially since in focusing on leadership we are considering several levels simultaneously—for example, from the designer to the first teacher who undertakes ChemCom in her class. Also, we are using the term "leadership" to designate the authors of key documents, the initiators or guiding spirits of the UMC collaboratives, and the science specialists who carry a promising telecommunications project (the Kids Network) through a lukewarm administration. Yet there are one or two pioneering figures (e.g., in the Precalculus project, Project 2061, and the California reform) who remain prominent. Perhaps the only common thread is a dynamic corps of teachers early on in a reform or innovation, acting as critics or linkers, or later on as staff developers, project coordinators, and operational leaders (in UMC, in particular). Elsewhere, these often were the people who assumed many of the functions of principals or even of district-level administrators.

The California reform covered all these bases: involving teachers early on in the design of regional or districtwide projects and creating differentiated leadership and staff development roles for teachers; these roles were akin to actual career shifts (lead teachers, staff developers, consulting teachers). There was

Table 7-5.   Leadership Roles in the Eight Innovations

Project	Primary Leadership Roles	Dilemmas
California reform	Central figure, but diffusion of leadership through networks of like-minded project leaders	Danger of exhausting hub coordinators and CSIN leaders, given the multiple roles as teacher-educators and in district-, school-level, and classroom work
Project 2061	Visionary at the top; more executing at other levels of project	One central locus of authorship, design, fundraising; yet problems of succession are real
NCTM standards	Shared leadership in drafting and dissemination, with researchers assuming quality control, but with strong initiatives in the central and regional professional associations	Balance between leaders at different levels; gradual handing over of some initiatives to districts, school leaders
UMC	Teachers taking on leadership roles in districts: presenters, grant writers, delegates to district administrations	Danger of role overload: administering collaboratives, keeping up with developments, keeping classroom hours
Precalculus	Central figure, surrounded by a team of core collaborators	Leadership construed as a collective function, but little desire to invest in achieving impacts in other schools
ChemCom	Initial direction from the professional society, then delegation to local levels	Tighter links between local leadership and chief sponsors in the professional society, but also unwelcome local transformations, distortions
Kids Network	Design leadership also given to publishers, local administrators, science/technology specialists	Problem of separation between developers, publishers, and local users: different perspectives for each party
Mimi	Design leadership handed to publishers, then to administrators and active teachers	Problem of separation between developers, publishers, and local users: different perspectives for each party

even targeted training for school administrators, through CSIN, combining substantive mastery with skills for developing, organizing, and furnishing ongoing support for program implementation.

A final word on the Precalculus project. The case study records less an adoption process than a development project, resembling in many ways the development process of Voyage of the Mimi at Bank Street and Kids Network at TERC. The case study portrays the inner workings of staff collectively pushing the envelope toward student-centered, inquiry-oriented mathematics and science. For many, in fact, this was a far more serendipitous process than in the other curriculum development projects—and one whose results often surprised its own developers as they experimented with graphing calculators and spreadsheets, introduced modeling to students with little guidance, used functions to represent real-life complexities, taught trigonometry without the use of the unit circle, and gradually introduced ideas of iteration that could lead into exponential functions. More generally, and not without hesitations or internal debates, the Precalculus team constructed a more functionalist approach at the expense of the traditional structuralist or formalist mathematics curriculum, learning from field trials as they worked their way to a textbook.

Teacher Background in Mathematics and Science

The dichotomy between process and content did not seem to pose a problem to the Precalculus team, whose grasp of the underlying mathematical content and conceptual frames was apparently solid. Nor was this an overriding concern at UMC collaboratives, which relied on teacher volunteers. In ChemCom, the problem was perceived more as lack of preparation for the unconventional parts of the course, with some chemistry teachers in particular avoiding "all these social science things" that ate up the time required for the purely chemical aspects of the curriculum.

It was, however, a leitmotif in the NCTM project, particularly at the elementary level. There were intuitions from the start that changes in the curriculum would not compensate for the shortage of teachers prepared to teach the recommended mathematics. And these teachers would require a wider and deeper repertoire than previously. Thus, professional development was more than a privileged, systemic component for several of the projects (e.g., the California reform, UMC, and the NCTM standards); it was a linchpin. Of what use were standards in mathematics if teachers were unable to derive the multiple tasks and representations allowing their students to reach them? Moreover, in several projects, California reform and UMC among them, instructional facilitators were not always better grounded than the teachers with whom they were working in the mathematics or science required to implement the reform or innovation. Whence then the provision of the scientific knowledge required to teach physical, earth, and life science? Whence the quality control in individual projects?

This was a constant in the elementary level projects—and a consequential one. Students' prior conceptual knowledge influences all aspects of their thinking and problem solving. Correctly gauging and interpreting levels of prior

knowledge and conceptualization on the part of the teacher was thus indispensable, and was inseparable from the teacher's own capacities in these areas. In
effect, optimal levels of discrepancy, challenge, and capability—and of pupils'
actual knowledge and skills—are mediated by the teacher's own understanding
of the deep content structure. It was no accident that the more highly trained
Precalculus teachers were able to resolve in large part the spurious nature of the
debate between the complementary roles of deep cognitive process and content
knowledge.

The Need. Teachers do not draw uniquely from their knowledge of subject matter, nor from the particulars of curricula; there are other requisite skills
in the creation of effective teaching strategies in the classroom. Above all, there
is the transposition of content knowledge into appropriate and motivating experiences—the principal source of student learning. But teachers' knowledge bases
do affect their decisions and actions in the classroom. Experienced teachers'
understanding of a domain is interwoven with their knowledge of how to teach
it. And practices are unlikely to change in the classroom without corresponding
expansion, enrichment, and elaboration of that knowledge base, including its
substance, explanatory framework, procedures, and their interconnections
(Grossman et al. 1989). For some researchers, in fact, knowledge of subject matter actually includes an understanding of particular topics, procedures, and concepts, as well as of the organizing structures and connections within a discipline
(Ball 1988). How, for example, to limit coverage yet ensure conceptual understanding? How to structure content around the powerful ideas rooted in the discipline, and in the relationship between those ideas? How to provide articulated
tasks, opportunities, and engagements in class that call for the conduct of
inquiry, the resolution of problems, or—more generally—the exercise of other
higher order applications of the content matter in a variety of situations? If the
process skills in student activities have to do with making observations, formulating hypotheses, controlling variables, elaborating models, etc., they have to be
anchored in the teacher's substantive grounding.

Thus, to implement the kinds of mathematics and science education reforms
intended by the innovations we studied, teachers needed to deepen their inherent
understandings of mathematics and science topics by actively building on their
prior knowledge. Participation with peers—a keystone of most projects (e.g.,
Mimi, ChemCom, Kids Network, UMC, California reform)—was probably
inadequate as the sole vehicle of mastery, although this is hard to assess. It certainly did no harm, but probably did not resolve fundamental pedagogical issues.
And it might have fed the delusion that child-centered science and mathematics
required greater clinical skills but perhaps not greater substantive mastery.

To extrapolate: In the coming cycles of implementation, one might need
more planful participation—and sometimes outright apprenticeships—by members of professional communities, through formal means (preservice and in-service education) or through more informal modes (networking mixed with training events). From the data of this study, it would seem that the reforms will

depend heavily on the substantive grounding practitioners can rely on as the reforms make their way into classrooms. Indeed, there is evidence that a likely consequence of inadequate grounding in the substantive areas of mathematics and science was the tendency to fall back on the familiar, reliable, more traditional formats for instruction. The cadre of teacher leaders, such as those studied in Project 2061, ChemCom, or California's CSIN may not suffice, especially if their roles slide more into the therapeutic than the didactic. Networking may be overly diffuse; even the intensive experience of participating in a UMC collaborative may be too short lived or too remote from the actual locus of classroom teaching to help many participants transform new curriculum plans into learning activities, which are the root source of students' learning.

Some Illustrations. The *California Science Framework* is an ambitious one. This set off the usual concerns: The *Framework* goals were unobtainable; there was too much expected of teachers; resources were inadequate; etc. At a second level, however, teachers complained of having too little experience and orientation in the direction of the *Framework*, i.e., in the conduct of meaningful and targeted activity in the classroom. In pedagogical terms, how was the *Framework* best approached at successive grade levels, across various bodies of knowledge and learning opportunities? What was important? Teachers were unsure of the changes in their instruction necessary to make it consistent with the *Framework*, and doubted that—even with this knowledge—they had the appropriate working conditions.

A unique characteristic of the UMC project in this regard was the building of a conversation around the links between content matter and transpositions into instruction. How to do this in settings beset with constraints (isolation, lack of administrative support, difficult students, scarce resources) and instructional unpredictability? How could substance, activities, and variability of contexts be built into an integrated discussion? The teachers who volunteered for a UMC project had opportunities to address these difficult questions. In the California reform project, on the other hand, there was, early on, uncertainty about the adequacy of teachers' scientific backgrounds and the lack of time to learn from what other teachers were doing in their classrooms.

With respect to Kids Network and Mimi, two further phenomena are worth mentioning. First was the tendency for teachers to fall back on the prescriptive activity guides for the choice of topics, conduct of experimental procedures, and discussion of results. This created the paradox of a supposedly investigative view of science and mathematics enacted in the classroom with predefined, standardized resource materials. Moreover, many teachers stayed with the materials for four to five years, taking few instructional risks yet gradually adding to their repertoire. Typically, teachers worked hard to learn the content; many added modest resource materials and sought out complementary experiences (e.g., speakers). Even aside from the technological components—for many, the most novel and unsettling aspect of both projects—few teachers profited fully from the potential of the programs. These were, after all, well- and expensively designed curricula. Some teachers, however, notably those with a stronger academic background, felt that the mathematics or science content was weak and was watered down further by some of the activities. As a Mimi teacher observed: "Much of the science is not really hard science. We're not talking physics and biology . . . We made signal flags. We had to do research, paper had to be cut, colors had to be filled in. But that's not pure science" (Middlebrooks et al. 1996, p. 446).

An interesting trend among Mimi teachers was to overcome their perceived limitations in science and mathematics by reading, through more careful planning, by attempts to devise some supplementary approaches, and through reliance on more experienced colleagues. Said one: "I probably spend more time planning science lessons than I do all the other subjects combined" (Middlebrooks et al., p. 444). Several excused their indifferent training in mathematics and science by noting that some course work was unavailable to women during their undergraduate years. Others took a more empirical stance, e.g., "I can do this activity and it's not going to be a disaster . . . and I can work this through with the kids" (Middlebrooks et al., p. 445). It *is* true that many Mimi and Kids Network teachers had a deeper understanding of the developmental aspects behind the reforms and innovations than protagonists in other projects, but that understanding was often used to attenuate the substantive (subject matter) demands. Others took refuge in the English or social studies aspects of the integrative innovations in Mimi, Kids Network, and ChemCom. And still others stayed focused on the canonical use of the materials prepared for them by developers and publishers.

The Role of Teachers' Beliefs

We have already commented on the "paradigm wars" and the continuum problem. We reemphasize here that there are good arguments for espousing alternatives or continuing with prior practices. One can be intelligently opposed to reforms. To some extent, these are political options.

Within the projects, there was certainly debate. Even within the Precalculus project, some teachers thought there had been too extreme a shift from pure to applied mathematics. Strong believers in a basic skills perspective were rare or short-lived participants in UMC. Nonbelievers in issues-oriented science left ChemCom. Some NCTM recruits among teachers saw the standards as providing too little computational mastery and felt their administrators saw things the same way. They had hoped for a more checklist format; some also had special interests in more challenging standards for their advanced students. But others felt that the standards were far too timid. At NCSSM, for example, some teachers had trouble seeing the merits of the NCTM *Standards*.

These differences in perspective come from a number of sources, including experience and training, role models, varying epistemological stances, and local policy choices. In the elementary schools, prior beliefs of teachers are hard to shake, since their source (personal experience as a student, brief training, arduous coping in the initial years) is so emotionally underpinned. Then, too, there were conservative students—as, for example, in the Precalculus project—for whom a more problem-centered approach to mathematics raised anxieties ("we'll be held back"), and others for whom interactive inquiry was a teaching style they disliked.

"Teachers' beliefs," then, represent an ill-defined territory. Strategically, however, what was crucial was a teacher's experience with or sensitivity to alternative paradigms. For example, the more dynamic teachers in the reforms and innovations often had a long personal and intellectual career history; some had trouble reaching back to the earlier stages they felt were represented by many of their colleagues. Like their counterparts in policy circles, they were in a hurry. And they might then have lost perspective on the paradigm they were advocating, as if there were no other. All projects seemed to feature individuals with these characteristics. In the curriculum development projects (Voyage of the Mimi, Kids Network, ChemCom, and Precalculus), teachers deliberately used the self-contained classroom to sustain their individual beliefs—often by screening out interlopers.

The Burden of Change

The links between reform policy and day-to-day implementation at the instructional level are best observed by studying the finer grained enactments of daily practices. An important aspect of daily practice is the actual burden of implementation, discussed in this section mainly from the perspective of those shouldering it.

Historically, innovations suffer from an overload of activity at the very time they require a more casual, spaced-out schedule for teachers to assimilate tasks and unexpected results. After all, innovating teachers are in the business of managing uncertainty, no matter how modest the mandate. Mandates for reform appear to be univocal; but they activate so many variables, both identifiable and expected ones as well as subterranean and as-yet-unseen invisible ones—all interconnected—that the capacity for loosely managing the setting is more important than the ability to get all the right pieces in place.

To make an exhaustive list of constraints faced by teachers and to interconnect them among projects is beyond the scope of this chapter. Below, we list and describe the most frequent burdens and note which projects were afflicted by them more markedly. We should note, too, that the reforms and innovations did not necessarily *create* these difficulties, but often provided a new theater for their emergence locally.

Conflicting Expectations. At the simplest level, the reforms and innovations were out of kilter with their surrounding environments. For example, parents in schools claiming to be aligned with the objectives of the NCTM *Standards* found that their third grader did not know what multiplication was, nor had their sixth grader worked with fractions and ratios. It took diplomacy to put their concerns to rest. Feeder schools at the sites observed in the NCTM case study were putting more pressure on teachers at higher levels—who were already overloaded. Some ChemCom teachers with a strong chemistry background went very gingerly toward the curriculum's social issues perspective, and not without skepticism. The Precalculus experimental schools also had staff and parents of varying orientations making different demands.

Overload. "Less," to paraphrase the Project 2061 dictum, became "much more." Interactive, problem-solving pedagogy is voracious in its demands for new materials, activities, and experiences. Learning through active experience cannot rely on the inventiveness of teachers and pupils, especially in the lower grades. Thus, the developers' decisions in Mimi and Kids Network to provide comprehensive kits appear quite reasonable. Staff development time also began to be monopolized with this concern, which in turn cut down or cut out a more contemplative or openly didactic exchange. The lack of a broad conversation about science reform in the California case might well have lowered the gradient of pedagogical change or the quality of the science it actually precipitated.

We note also that as projects became more autonomous, they took on more responsibilities: coordination, quality control, continual choices between lesser evils, urgencies, recalcitrant peers who were not like-minded or whose expertise lay in more "traditional" science, etc. Clearly, these roles entail levels of commitment that could not be sustained unless they were rotated (as, to some extent, in the Precalculus project) or codified.

Infrastructure and Social or Organizational Constraints. Kids Network was obliged to work with outdated or unreliable telecommunications. Also, to meet the publisher's time lines, teachers were forced to rush through the curriculum, leaving no time to process the community database; debate the findings; and shift into a more interpretive, reflective mode that was meant to constitute the curriculum in the first place. Mimi teachers had to find ever-more ingenious ways of hoarding the few computers available. The UMC teachers left their meetings ready, for example, to experiment with graphing calculators, and then were confronted by algebra requirements on standardized tests, hard-to-reach students, inadequate resources, and indifferent—when not outright hostile—department colleagues. Project 2061, UMC, and the California reform were distracted by the quest for long-term support. ChemCom (see below) contended with unwanted shifts in programming and population. NCTM continually ran the risk of being no more than the sum of its constituents, and thus something of a mathematics policy football for the education reform and anti-reform advocates.

These examples illustrate a paradigm outrunning its environment and threatening to wear out its most committed protagonists. They remind us that there is no one "out there" to oversee the process. Nor are there stable protectors at the head of crucial agencies and commissions. This creates a need to vary the pace, to live philosophically with setbacks, to act opportunistically and take advantage of openings, to set up loose monitoring mechanisms, and—generally—to live with uncertainty and improvisation until the remaining pieces are put, imperfectly, in place. Overnight, in fact, a stronger set of determinants can carry a reform or innovation in another direction.

A clearly observable effect of phases of change in the projects under study was a gradual or temporary backsliding, sometimes one imposed from outside: slowing down the pace or reducing the increment of change (NCTM standards, California reform, Kids Network); consolidating a smaller number of key nodes (UMC); making alternative arrangements (the relegation of ChemCom units to high school remedial science in the 10th and 11th grades). Note also the drift of Mimi toward a more purely vicarious science, with few actual experiments or experiences for the students.

Users of Precalculus, ChemCom, Mimi, and Kids Network may well have the kernel of a strategy for survival and growth: Work with already developed, field tested curricula that meet local criteria of purpose, quality, and orientation; then reinvent the products locally. For developers and teachers in larger scale projects, writing new curricula in times of shrinking budgets (California reform) was, perhaps, "grassroots suicide." To paraphrase the CSIN evaluator for California, "developing really good curriculum material and giving that to teachers . . . might be the most cost-effective approach . . . It's an open question" (Atkin et al. 1996c, p. 111). This is, in many ways, the implicit strategy in some projects—e.g., NCTM standards and Project 2061, with their progressively more operational guidelines. But these publications are not curriculum material in the same sense as the Precalculus textbook. It may well be that the next task for the larger scale reforms will be to produce some prototype materials or identify which of the smaller scale projects or applications are consistent with their approaches, as Project 2061 is currently doing.

But this is not the only way to construe the problem. Consider poor organizational fit between the demands of the innovations and the host settings. For now it appears to be the settings that dictate the goodness of fit. Either there is compliance or there is adulteration of the innovation. ChemCom teachers, for example, began doing the integrative science modules in parallel to the existing curriculum; or, alternatively, fitted in more subject-specific content for chemistry majors. ChemCom also became the basis for 9th grade integrated honors courses, putting it at odds with one of its main objectives: to attract students not interested in the traditional 11th grade chemistry course. Potential 11th grade ChemCom students liked the environmental focus of the modules, but pulled out when they saw that academically oriented peers were not enrolled.

Some experimental schools in the Precalculus project found the textbook too thin and requested supplements; others found it too ample and wanted a slimmed-down edition. Some of the teachers ran into demands that their perspective be consistent with the canonical Advanced Placement (AP) syllabus in calculus. In California, with new frameworks enacted in science periodically, the

1990 *Science Framework* had to build in plasticity, especially since science was already considered an add-on or elective. More emblematically, at one of the NCTM standards sites, student tables went back into rows, and a more disciplinary emphasis made teachers pick selectively from the process approaches implied in the *Standards*.

Poor organizational fit, however, has merits as well. For example, the *California Framework* strengthened the hand of science educators and teachers who felt they now had a mandate to change local practices. In other words, the reforms legitimated changes in local practices where there were adequate degrees of freedom to change. This was the case as well, but with more friction, for some schools struggling with the NCTM *Standards*.

Aside from legitimating, poor fit indicates where the reforms most likely will run up against existing organizational structures and working arrangements—not to mention informal loops of power and influential gatekeepers who can kill or accelerate new initiatives, distribution of resources, curriculum and testing regulations, powerful external constituencies, changing characteristics and attitudes of student groups, variations in student performances, teacher backgrounds, and so on. Information on organizational fit is valuable. For one thing, it can help set the agenda for the next, more operational, phase of a reform by identifying likely zones of friction at the institutional level. For another, as an intruder in a new surround, one of the prime roles of a reform and its attendant innovative practices is actually to create discrepancy, especially at the instructional level. For example, some of the potential discrepancies in Mimi had to do with new science content, more hands-on activities, experimentation with cooperative learning, integrated technologies, interdisciplinary curricula, and instances of a more prominent role for women and minorities. Some innovations are even potential institutional change-bearers; for example, regarding the local mathematics or science curriculum or when scheduling and deployment of staff are at stake.

Without such discrepancies, very little will change. With too much, the innovation will generate more turbulence than the institution can handle. There is thus an ongoing, often implicit, battle for influence between the change-bearing demands of these projects and the equilibrium-preserving arrangements by which schools manage so many moving parts. Which will prevail? Unless the innovation is limited to within-classroom activity, host schools will not spontaneously modify the conditions of learning that, in fact, seem to deliver significant learning outcomes. To implement forcefully is to court temporary disruption and—not always so temporary—conflict. As a result, the typical journey of a new program through the institution results in phases of watering down—or "domesticating"—the most change-bearing components. The process of local appropriation is thus a dialectical one: often painful, uneven, uncertain, even periodically "unmanageable," yet the best route to significant change when the requisite accommodations are made to the demands of the reform or innovation.

By so altering the operating environment, the school creates the opportunity to alter its instructional processes in significant ways.

Weick's (1984) construct of "small wins" may be pertinent in this context. In both the comprehensive reforms and smaller scale innovations, we observed people defining the problems of implementation in ways that overwhelmed their ability to do anything about them. By contrast, a small win (e.g., reinstating segments of the ChemCom curriculum to counteract creeping "mathematization" or applying the NCTM *Standards* concretely to a small set of participating schools) typically put forces in motion that favored another small win. The next solvable problem often became more visible, bringing new, as-yet-unseen solutions—and often attracting slightly more resources. This set the stage for slightly larger wins. Globally, the particle cloud strategy exemplified by the California reform worked this way.

Weick insists that such attempts be specific, realizable, and immediate. Small wins stir up settings such that they cannot be carefully plotted in continually changing conditions. "They do not combine in a neat, linear, serial form, each closer to some pre-determined goal . . . [Rather] small wins are scattered and cohere only in the sense that they move in the same general direction . . ." (Weick 1984, p. 43).

Small wins may well accumulate to create the conditions for major change as suggested by Fullan (1994). Drawing on Beer's later work, Fullan sees a sequence whereby teachers and administrators begin working in new ways, discovering gradually that school structures must be altered (Fullan 1994, p. 194). The pressure mounts to change organizational arrangements that are now experienced as ill-fitted to the emerging patterns. Formal procedures and formal reorganization then begin to kick in, but later in the process (Fullan 1994, p. 198). At this point, they would presumably join up with the broader reforms being put in place "above" them, such as those depicted here, creating simultaneous pressures from inside and outside.

Changes in Teacher Capacity

The literature on innovation mastery at the individual level (e.g., Hall et al. 1975, Huberman and Miles 1984) is fairly robust. It replicates earlier, classic work and intersects well with the data available from our case studies. In the UMC case, in fact, there is a model depicting the trajectory of teachers through their association with the collaboratives and with similar modes of professional development (see below).

In Hall's (1977) work on innovations, adopters' initial concerns tend to be personal ones having to do with teachers' own personal or professional adequacy—how the innovation may affect their status, their sense of professional competence, their rewards. These concerns change gradually to concerns about the correct execution of the innovation and the ensuing impact on students, and then to concerns about whether other innovations or a new configuration can achieve the same goals more effectively. The process varies somewhat by teacher; for example, the degree and quality of assistance received can change the profile, as can a benign infrastructure (favorable class assignments, good materials, supple regulations).

The teachers observed in our case studies appear to follow the general pattern. They seem to move progressively from personal concerns to technical mastery (e.g., computer modules), then to matters having to do more with the program's impact. This triggers a search for ways of extending or deepening the program.

Initial Phases. Mimi and Kids Network serve as examples. In almost all cases, regardless of the teacher's previous background in science, years of teaching, grade level, availability of or experience with technology, Mimi teachers characterized their first phase of use as feeling both excited and overwhelmed. The latter sentiment had to do with learning and teaching unfamiliar content and how to teach it.

These sentiments went accompanied by complaints of day-to-day coping, unsuccessful attempts to "make it work like it's supposed to", successive cycles of trial and error, periodic exhaustion getting through daily or weekly segments, and the sacrifice of time for other core activities (e.g., segments of the traditional mathematics or science curriculum). In their own way, however, difficulties at the outset were good harbingers. They signified that teachers were genuinely trying to face the discrepancies between their own practices and the change-bearing features of the new program.

The difficulties appear related to the overflow of seemingly simultaneous tasks ("so much coming at me"); unpredictability ("sometimes an experiment works; sometimes it doesn't, and I did the same things"); and a lack of understanding as to how the overall program is constructed and interrelated. There is, then, both a mental and a technical task at each successive phase of the process.

Later Phases. The progression might best be rendered by a simplified figure of the implementation process. Applied to such projects as Mimi, Chem-Com, and Kids Network, figure 7-1 is an illustrative simplification. It underrepresents the nonlinear quality of practice mastery: the plateaus, regressions, sudden spurts, long moments of latency, etc. Generally, however, each of the steps denotes a different phase in the implementation process. The phases are largely self-explanatory, but a key might be helpful.

"Stepwise use" implies disjointed activity, which is resolved in part by following the project guidebooks slavishly. This was, in effect, a feature common to the curriculum-based cases. "Initial coordination," or routine use, calls for few or no changes: The teachers have the program's rudiments pretty much down and stick with them. "Coordinated practice" involves teachers making some changes to increase the program's impact on students, typically supplements or small variations. Finally, a "refinement" phase entails an active search for alternatives, parts of which are meshed with the new program. Classic (e.g., Hall and Loucks-Horsley 1978)and more recent (Huberman and Miles 1984)

Figure 7-1.   The Classic S-Curve

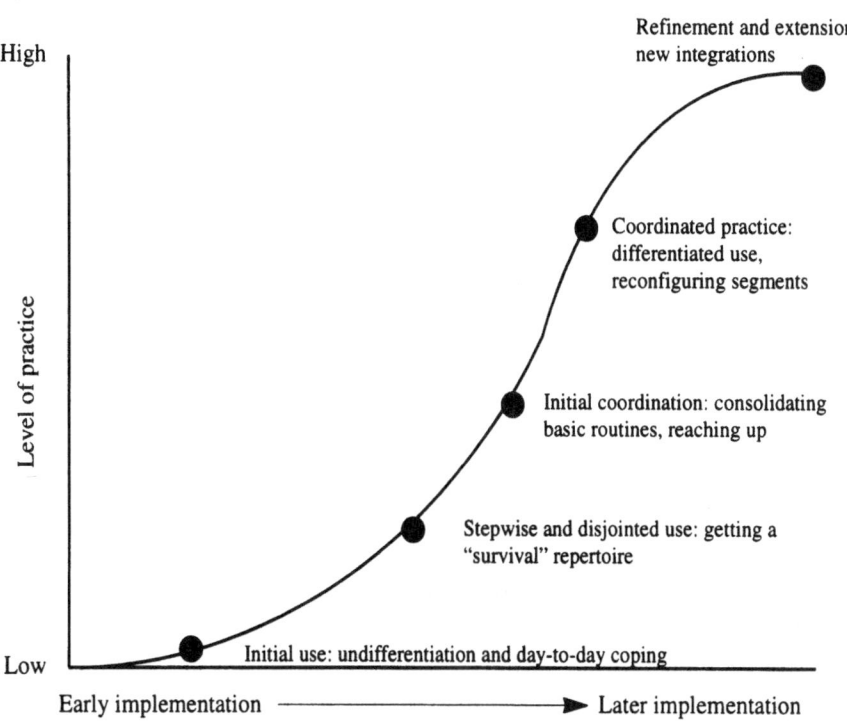

Refinement and extension: new integrations

High

Coordinated practice: differentiated use, reconfiguring segments

Level of practice

Initial coordination: consolidating basic routines, reaching up

Stepwise and disjointed use: getting a "survival" repertoire

Initial use: undifferentiation and day-to-day coping

Low

Early implementation —————————▶ Later implementation

studies found that it took a good three to four years before teachers began to get beyond the "routine" phase. One noteworthy feature of the full-cast model (Huberman and Miles 1984, p. 125) is that it integrates cognitive progressions (moving from "undifferentiated" understanding to "integrated" understanding) and technical expertise, so that more coordinated understanding interacts with more extended mastery.

It typically took Mimi teachers two to three years to reach a phase of initial coordination, but a few were already at succeeding phases and were beginning to combine Mimi with other—sometimes new—components of their instructional repertoire. One or two reached the refinement-extension phase. Virtually all teachers, however, remained at least partially in an algorithmic mode, as in the earlier stages of the figure, in which pupils followed a series of largely predetermined steps to the correct answers. Nevertheless, much of this was hands-on science and mathematics—actual hypothesis testing, derivation, inference, marshaling of evidence—and, as such, represented a new approach for many teachers in the sample.

Modifications in the Precalculus project came earlier than in many other innovations, in large part through ongoing experimentation. Case analysts in this project were struck by how often participating teachers reported making changes in units. They felt comfortable modifying, reordering, or dropping units that did not seem to work as they should. One teacher, for example, moved the modeling section to the end of the year as a final project rather than treating it in class. Another incorporated a unit on statistics that was not in the Precalculus text-book. One might contrast this example with Mimi and Kids Network, where par-ticipating teachers—mostly orphaned from the outset, with little preparation or contacts with developers or trainers—saw the teacher's guide as a sort of life-line.[2]

Conceptual progressions are well-captured in the UMC case. The authors provide data showing how beliefs in computation and algorithms gave way to understandings of processes and of interconnections among new mathematical topics. At the same time, actual applications in frequently difficult settings added the more technical components, as in figure 7-1. Teachers ran experiments lasting days to weeks—e.g., using graphing calculators to process data, learning about parabolas and tangent lines by throwing balls into trashcans, plotting growth rates with dehydrated dinosaur toys. One of the keys here was the active encouragement of risk taking, which in turn transformed learning environments for pupils. Teachers implementing the NCTM *Standards*, for example, seem to have been far more cautious. So were the science educators associated with CSIN in California, when they were obliged to turn to the elementary school teachers to flesh out curricular substance.

While they were closer knit on conceptual issues, staff at the Precalculus project interwove continuous experimentation ("It was a full year of just trying things out.") with ongoing debate and new trials ("If the idea was not successful, it was thrown out.") (Kilpatrick et al. 1996, pp. 153-54). Only gradually, and in large part by having to codify their ideas in the text, did they come to a better grasp of what they called the "big picture" of what they were trying to do, the "big ideas" of the course—interpreting data, developing mathematical represen-tations, transforming and using functions—and how students could make sense of them. Here again, understanding was achieved, not derived in advance.

Enactments: A Closer Look

Evidence of progressions, even though some of these are due to the assistance furnished to local actors (see the section on "Assistance" below), suggests that

- significant changes were enacted at the school and classroom levels; and

- these changes were affected by, or derived from, the reforms or innovations that were undertaken.

What evidence is there for these claims? Why suppose, for example, that the availability of the NCTM *Standards* or the Project 2061 *Benchmarks* would

---

[2]One difference is the teachers' subject matter background: strong in the case of the Precal-culus teachers, which allowed them to modify the units; weak in the case of Mimi and Kids Network, which made them adhere closely to the projects' materials.

change teachers' conceptions or practices? The NCTM case study itself, after a tour of class-level activity, concludes that there was little implementation in the classrooms observed, even among capable and active teachers. The reasons given apparently had to do with teachers' conservative beliefs and with the constraints of daily life in the classroom and the school at large.

Case researchers found that the *California Science Framework* did not initiate reform activity, but rather amplified or legitimated reform-oriented work that was already under way. Enactment here had more to do with putting in place the infrastructure of the reform (staff development, assessment, integrated curricula) than with executing it on the ground. Still, there were testimonies of follow through. CSIN, for example, with its focus on whole-school reform, reported some degree of implementation in selected districts. There was measurable movement, also, in the trial of new, performance-centered assessments. Distinctions between "experiential" and "constructivist" science became sharper at the classroom level, not just in forums. Most important, teachers actually came to grips with the conduct of integrative science, collaborative inquiry, and construction of open-ended curriculum materials meant to replace "cookbook labs." Large numbers of teachers experimented with relevant strategies in their classrooms. The new paradigm in science education got tried out, possibly before its time, even by teachers working outside of experimental settings. Overall, however, the California case study suggests sporadic, high-mortality implementation—but *some* implementation all the same. In California, the impression was that of cautious initiation in classrooms, with periodic backsliding when the change gradient got too steep. Protagonists felt their way slowly, electively, irregularly; but did not cause waves of criticism or calls for abandon, except for the assessment component.

Project 2061 school districts seemed to underscore their independence from the project's headquarters staff in Washington. Some continued on the course they had taken before their association with the project. The task of filling in the middle ground—between policy and benchmarks and between benchmarks and the kinds of learning opportunities that can achieve them—remains. With ChemCom, one observes an innovation widely adopted, but struggling to avoid slippage; as teachers supplemented the curriculum with other materials (for example, excerpts of *Chemstudy*, a text based on the 1960s reforms), failed to pursue some of the social questions, tried to make the mathematics more rigorous, and abandoned some labs because they found the choices of materials or quantities inappropriate.

Mimi Enactments

We provide a more detailed look at the teaching of Mimi, based on 135 hours of observations. We expected Mimi teachers in their fifth or sixth year of use to wean themselves from the teacher's guide. Yet the modal enactments were scripted, teacher-directed activities, as illustrated by the two following examples.

"Faithful" Use of Mimi. In this enactment, the teacher closely follows Mimi's overview guide. Typically, the showing of an episode or expedition is preceded by preview questions and then followed by selected follow-up questions.

Figure 7-2.   Faithful Use

1. Teacher poses several questions to class ⟶ 2. Class watches a Mimi episode
3. Teacher poses several follow-up questions. ⟶ 4. Students work alone or in pairs on a predesignated assignment.

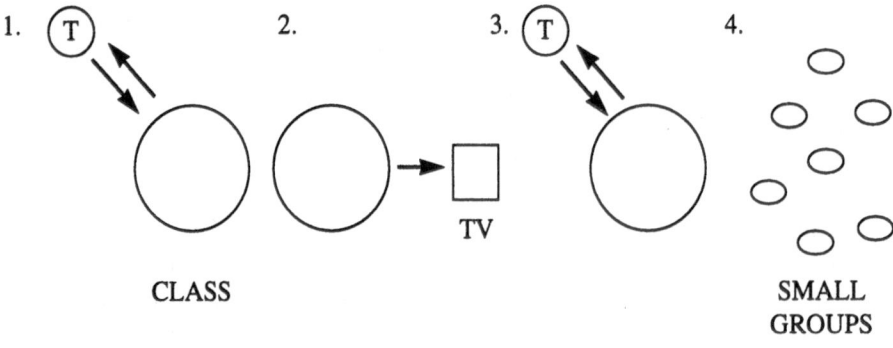

As observed by Middlebrooks et al. (1996, pp. 463-64), the two third grade classes of 40 pupils gather on the rug in HP's room at Site 1; the three adults (teacher HP, student teacher KC, and the paraprofessional who works with children who have special needs) sit or stand around the edge of the group. The children ignore the large TV in one corner and face KC as she asks questions suggested in Mimi's *Overview Guide for Expedition Five*, related to animal adaptation, and gives a brief summary. When the TV is turned on, the children hum to the theme music that accompanies shots of the boat under sail. The children stop as soon as the Mimi character Rachel explains that she will be visiting the Marine Biological Laboratory at Woods Hole, Massachusetts, to learn how marine animals are collected and used for research purposes.

After the show—and more humming—KC continues with more questions from the guide. She then goes over several vocabulary words, including "chemoreceptor," and suggests that, at lunch, students try holding their noses while eating their sandwiches. (This "experiment" is her addition.) Probably because a horseshoe crab is featured in the film, one boy interjects, "I know why the horseshoe crab got its name," adding that he has a shell at home. (KC does not suggest he bring his specimen to school.)

Toward the end of the period, KC asks the group for their "thoughts and feelings" regarding the removal of animals from their habitats for the purpose of research and experimentation. Unlike other questions, this one is not from the overview guide. Children's responses are diverse, and many seem eager to voice an opinion:

- "I wouldn't want to experiment on an animal."

- "I wouldn't want to be that animal experimented on."

- "If we don't do it, we don't find out things that are helpful to us."

- "Even if an animal returned to its habitat, I don't think it will find its family again."

KC then wraps up: "We get information from experimenting on animals and that helps us, but we shouldn't take more than is necessary." The two classes separate. The students remaining in HP's class soon are using dictionaries to look up the short list of words they have been given and to write out their definitions. They work in pairs with friends. Later, in conversation, HP speaks of the activity as a double lesson: With a simple activity, children can learn scientific vocabulary as well as how to use a dictionary.

During this observation, KC asks the questions, staying close to suggestions in the Mimi guide. In this way, the teacher keeps the pace of the discussion moving and chooses what is "important." Students give short answers, and a list of topics and concepts gets "covered."

Modified Use and Extensions of Mimi. This enactment is an example of modified use, but actually a variation on the faithful user scenario. In step 4 of figure 7-3, the culmination of a project was observed that teacher PW calls "marine mammal reports," namely students revising their written reports, making a diagram or a paper cut-out, or otherwise rendering their mammal through writing, speaking, drawing, modeling, etc.

As Middlebrooks et al. (1996, pp. 473-74) observed, PW returns drafts of students' written reports (each was checked by another student, then by PW). At 9:50 a.m., the students put their homework away, and PF reviews how to do oral reports, as students sit with notes and their marine mammal models made from large sheets of black-and-gray paper. PW goes to a large world map hanging in the front of the room to show a boy where his whale lives; another boy joins them.

Reports begin. The protocol is to stand before the class; talk for several minutes; show, for example, the paper model; and take three questions from the class. Despite several interruptions (by adults and students from other classes), the young speakers give clear, coherent, and smooth recitals. (Some children later report that they had followed their teacher's advice and practiced the night before either in front of a family member, the mirror, or a toy animal.) Class members appear to listen attentively to each report, but their clapping declines as time goes on. The girl reporting on the narwhal is asked questions, but several students are not asked any.

Figure 7-3.   Modifed Use — Extensions

1. -2. Mimi episodes and expeditions shown ——▶ 3. "Projects" follow with students working individually, in pairs, and in small groups ——▶ 4. Students present their work.

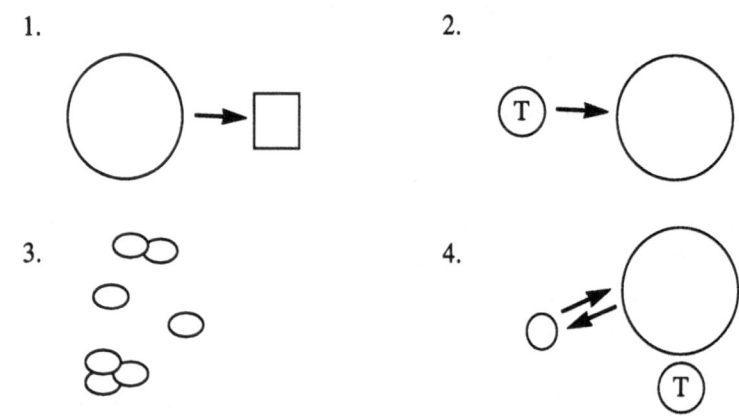

As the culminating activity, PW asks each student to write a summary of what he or she has learned. Her purpose, as she explains later, is to elicit and then assess each individual's experience. She tells the students to work alone. A special sheet of paper, lined with whale flukes at the top, is handed out for this purpose.

These enactments were not, as we have noted, the original intentions of Mimi's developers. Mimi's introductory materials invited teachers to be adventurous with the overview guide and student book, and to self-design activities that teach for understanding. The guides, together with the television series, suggest no concrete models of ways teachers might have organized classroom discourse around powerful ideas, experimentation, and the relationship between the two.

The Kids Network

This curriculum involved more actual experimentation then did Mimi, but, as noted, the topics, problems, and procedures were mainly selected from the guides by participating teachers. Observers also noticed early on that teachers were spending more time managing activities and materials than probing students. Was the modem connected? Were resource materials out and available? Despite wide use of terms such as "hands-on science" or "inquiry science," observed practices contained few examples. Teachers seemed more concerned with science process skills (e.g., theorizing, controlling variables) than with the understanding of scientific concepts (e.g., the effects of pressure on weather conditions).

Still, as prestructured as these investigations were, from a pupil's perspective they could be adventurous. They contained unknowns and obstacles: interpreting the directions accurately, setting up the materials, recognizing group or individual discrepancies, and reconciling differences from telecommunicated data. Furthermore, students sometimes generated new ideas from their participation with standard experiments. Finally, teachers who took liberties with the guides could take these exercises directly into a more open-ended inquiry mode, as the following example of a student-initiated chlorine problem—making a "yeast monster"—demonstrates. (See Karlan et al., pp. 328-29.)

Every student follows the same set of directions to determine the effects of chlorine on yeast reproduction. At the end of this activity, a sixth grade boy suggests that they make a yeast monster. With KA's enthusiastic support, they begin. Students immediately rise from their chairs; while smiling and laughing, they begin pouring all of the remaining molasses into one container. The class gathers around the center table to mix a large batch of yeast, water, and molasses. They initially want to mix all they have left from the previous activity. KA suggests that they keep the proportions the same as in their smaller cups. She asks the students to multiply by 3 the amount of molasses they used in one of their chlorine cups, which was 2 1/4 teaspoons. One boy calls out almost immediately that they will need to add 6 3/4 teaspoons. The teaching intern asks how he arrived at that number. He explains that he multiplied the 2 by 3 and got 6, and then multiplied the 1/4 by 3 and got 3/4. While KA mixes the ingredients, there is a constant buzzing of suggestions and predictions.[3]

Students derive a variety of concepts from their yeast experiments; some intended by the program, some not. For instance, one fifth grade boy summarizes: "We found out that chemicals kill yeast wicked easy because we had one with chlorine in it and it was about that big and the other one was huge and the yeast grew." Three other boys, high-performing science students in the class, explain during a group interview that the yeast was a microorganism that began as a dry hollow seed. In their view, the yeast grew into large bubbles on the surface of the water. (They misinterpret the evidence for the presence of yeast growth—the release of carbon dioxide bubbles—for the yeast itself.)

The Precalculus Project

It is difficult to determine whether the enactment of the Precalculus project is its textbook, together with the continuous experimentation preceding it, or whether one needs to look inside the five schools observed by the case study researchers, where the Precalculus textbook was diversely appreciated. At Woodward

---

[3]Another boy offers an alternative strategy; he converted the mixed number, "2 1/4," into a fraction. He multiplied the 4 (the denominator in 1/4) by 2 and then added one (the numerator) and got 9/4. He then multiplied the 9 by 3 and got 27/4. Before this boy finishes his explanation, KA indicates that his method isn't right. Of course, it produces the same result as the first boy's strategy.

Academy, observers found concerns about looming AP curricula and tests, students clamoring for more textbook examples, teachers split on the thoroughness of the textbook, and parents nervous about content mastery of traditional mathematics. But several teachers appreciated the inductiveness of the textbook, although they may have been headed that way on their own. For the most part, Woodward Academy appeared to be an experimentally oriented school, with many teachers already subscribing to an investigative, problem-centered pedagogy. Even though the status of the textbook at Woodward may have been precarious, the enactment of an interactive, functionalist approach to mathematics was close to the ideal versions projected by NCSSM faculty.

Kilpatrick et al. (1996, pp. 179-80) record the following observation. As Paul Myers's third period class enters the room, they find a cup of water, a thermometer, and a stopwatch on each table. As class begins, Paul explains that the students are going to conduct an experiment and record the results. The experiment involves recording the temperature of the water every 10 seconds after ice is put in it.

The students are asked to sketch their prediction of what the graph will look like. They begin to collect data. Paul instructs them to take an initial temperature reading of their water. He then places a handful of crushed ice in the cup on each table. The students work in groups of three, with one running the stopwatch and announcing when it is time to read the temperature, another reading the temperature, and the third recording it.

After all groups finish recording their data, Paul instructs them to draw a scatterplot on their TI-82 calculators. He then asks the class what their scatterplots look like. Jennifer notes that her graph "goes back up," and Paul asks why. She responds that eventually the ice will melt, and the water will return to room temperature. Lester says, "My graph looks like it has an asymptote."

Paul sketches a graph on the board and asks if the students' graphs look something like his. They nod. He asks the class if it is reasonable for the graph to have an asymptote. After some discussion, he asks what kind of functions have asymptotes. Several students respond that rational functions have asymptotes and are of the form $f(x) = k/x$. Paul asks if there are any other types of functions that have asymptotes, and a student replies that square root functions and exponential functions have asymptotes. Others argue that this is not the case. The students finally agree that square root functions do not have an asymptote and that exponential functions do. Paul asks what kind of exponential functions look like the one he has drawn on the board. A student says, "Negative, like a^{-x}." Paul writes $g(x) = a^{-x}$ on the board and says, "If you think this might be exponential, how could you check?" A student replies, "Re-express it."

Paul suggests that students re-express their data and look at the residuals. He reviews where actual values and predicted values go in the calculator and what to do with the re-expressed values. He wanders around the room, picking up students' calculators, examining the displays, and helping the students with problems and questions. A student is having difficulty (his graph runs above all of the data points) and asks for assistance. Paul picks up the student's calculator and says, "Tell me about your lists" (meaning tell me how and where you put your data and what you did with your entries). Paul questions the student until he figures out his own mistake and is able to correct it on his own. The period ends with a class discussion in which the group concludes that the students whose actual temperature readings were close to 0° C at the end of data collection had the best-fitting curves.

The UMC Project

The UMC project case study makes strong claims for changes in teachers' attitudes and practices—much of these derived from field visits—but provides little by way of actual enactment data. Aggregated, the data suggest that the collaboratives have been responsible for exposing teachers to new ideas, the latest thinking about how students learn mathematics, and what mathematics should be in the curriculum. The case researchers also have furnished explicit examples of how these ideas are being applied and of collegial help at crucial moments of actual experimentation in class. Teachers and students have apparently experienced an increase in enthusiasm toward mathematics in schools, particularly in cases of students who had perceived themselves as unable to do well in mathematics.

Assistance

What are the various sources of potential internal and external assistance, and how might they affect implementation? In table 7-6, we list some illustrative sources and their documented effects in our eight case studies, leaving out some of the most obvious sources of a more material nature (materials, equipment, direct funding).

As nonexhaustive as this list may be, it is still multilayered: aid from content matter specialists and district- or state-level personnel, from peers taking on staff development functions, from trainers or authors associated directly with program development; mobilization of district-level support and expertise; encouragement at the building level, furnished by principals or among peers. This assistance also takes multiple forms, both formal (orchestrated staff development work) and informal (networking and more assistance on demand). There is also recognition of the needs of different levels: aid to teachers, aid to trainers, aid to science educators, and, when they take on the teaching role, aid to researchers. Add to this the nature of the assistance and the variety multiplies: technical help, direct training, observation, discussion and debate, emotional support, methodological critique, conceptual clarification and scaffolding, administrative buffering and rule bending, and logistical support.

Whether assistance was adequate, well-timed, and competent was a different issue in each case. Well-supported projects, like ChemCom, did not necessarily

Table 7-6.   Illustrative Types of Internal and External Assistance and Their Enabling Effects

Project	Types of Assistance	Enabling Effects
California reform	Training for site-level administrators	Support, substantive competence for program implementation
	Staff development through CSIN, SS&C	Coordination in hubs; more direct assistance to schools via teacher consultants, lead teachers
Project 2061	Substantive assistance from teams of scientists and science educators	Elaboration of policy documents; elaboration of benchmarks and staff development needs
	Policy statements from state supervisors	Legitimation, reinforcement of network from policy to local-level administrators and educators
NCTM standards	Drafting, feedback, final development by high-quality teams	Increased credibility, ownership of document
	Availability of successive documents for "operationalizing" the reform (assessment, instructional possibilities)	Closer alignment of the standards with actual classroom settings; links between objectives and instructional activities
UMC	Collective vision of alternative mathematics	Conceptualization or legitimization of more constructivist perspective; sense of belonging to pedagogical and personal community
	Technical, social/ emotional, interactive aid	Greater risk-taking, experimentation; observation of exemplary practice, resistance to constraints and isolation in host school
Precalculus	Strong leadership, high level of individual, group competence	Forging of common goals; substantively rich collegiality through rigorous experimentation
	Explicit mandate from policy level	Possibility of full experimentation; degrees of freedom from local constraints (schools, parents) until textbook had been elaborated

Table 7-6.   **Illustrative Types of Internal and External Assistance and Their Enabling Effects (continued)**

Project	Types of Assistance	Enabling Effects
ChemCom	Development work by team of specialists and educators	Workable model for teaching socially oriented chemistry to nonmajors
	Initial and ongoing training	High level of preparation, on-going improvement of conceptual and technical skills of teachers and teacher-trainers
Kids Network	Strong support from principals: promotion of program, trouble-shooting	"Buffering" from administrative level, easing sense of non-pre paredness, uncertainties about content vs. process issue
	Technical aid from science and technical specialists (telecommu-nications)	Gradual mastery and inclusion or replacement of activities, materials; greater between-colleague assistance and sharing of computer-related problems
Mimi	Strong backing from principals; bending of institutional rules and constraints	Facilitation of interdisciplinary work; access to expertise and sources of technical training (e.g., computer use)
	Internal assistance: strong peer coaching and mutual advice, technical aid, nurturance	Encouragement of initial mastery in the absence of external aid; increased use of "hands-on" activity; gradual creation of school-level community in support of Mimi program

fare better than ones relying on volunteerism, such as UMC, given their very different objectives. The necessity of bootstrapping (Mimi, Kids Network) seems to have been beneficial in increasing teachers' commitment. In contrast, without long-term support, Project 2061 could not have put its successive pieces in place, nor could NCTM have followed its strategy of articulated documents. In the case of the California reform, a cloud of small-scale ventures, fairly continuous support, the creation of new materials, and follow through with larger scale organizational entities makes for an assistance-rich project, with emphasis on increased professionalism as a key component of the reform.

Student Perceptions

The case study material sheds some light on students' perceptions of the new curricula. As noted, some Precalculus students wanted more step-by-step approaches and feared for the AP Calculus to come. Students good at problem finding liked ChemCom, whereas students good at problem solving preferred standard chemistry. The majority reported greater sensitivity to social issues in science—for example, regarding power plants—and many found the program "fun and approachable," including the mathematics components.

Asked to consider how they might be acting like Mimi's scientists in their classrooms, students responded positively when they (1) studied the work of scientists; (2) watched the experiments and then talked about them as a class, posing hypotheses, for example; and (3) replicated experiments from the Mimi episodes. Nevertheless, doing science was experienced as "second hand" by students. They watched Mimi characters observe whales or, in the second voyage, listened to others' theories about Mayan culture. In the expeditions, students saw the labs and workplaces of scientists and heard about others' work. Infrequently, students replicated what they saw the Mimi characters doing—but in simpler versions.

Similarly, Kids Network students wanted more experimenting and "doing," as opposed to writing out material from texts and answering questions. Most highlighted the telecommunications component as one of the best parts of their Kids Network experience. Also appreciated was the work in resolving real-world problems, notably among students doing the Acid Rain and What's in Our Trash? units. Students enjoyed testing acid levels, measuring trash, recording cloud types and taking temperatures; then comparing their measurements with teammates and teams in other schools. However, students wanted more investigative activities (e.g., studying a landfill), more field trips, and more social action (e.g., setting up a recycling center).

Stabilization and Continuation

In 1978, Yin and his associates made an important discovery in their study of public service innovations: Successful innovations, even in terms of hard measures of success, could disappear within four to five years for lack of adequate "routinization." By routinization, Yin was referring to the inclusion of the project in cycles or yearly provisions for budgets, training, and personnel. Even spectacularly successful projects were no longer around five years later because too little attention had been given to their longer term survival, whereas some projects of doubtful value were well-embedded in the fabric of the county, district, or city.

The data on institutionalization are incomplete for the eight projects we studied, but they are suggestive. Some of the factors we consider important are laid out for each in table 7-7, which reflects informed guesses. At the time of the

Table 7-7.   Institutionalization and Continuation of the Eight Reforms and Innovations

Project	Likelihood of Continuation	Factors Favoring Continuation	Factors Threatening Continuation
California reform	++	Multiplicity of initiatives	Overreaching; necessary slowing of pace
		Fruits of networks already felt at district, school levels	Leadership instability; lack of proactive coordination of activity, resources
Project 2061	+/-	Mobilization of science community	Long-term funding
		Likely continuity of funding	Leadership succession, continuity of support from key constituencies, e.g., science supervisors, scientific community
NCTM standards	+	Standards have become reference point both nationally and at state levels; dissemination to districts, slow penetration of schools	Long-term support, resources
			Semi-political nature of the enterprise controversial among academic mathematicians
			Double-sidedness: challenging status quo, yet also vehicle of accountability
UMC	+/-	Commitment of leaders; widening of recruitment, e.g., in new areas	Lack of institutional base
			Financial, political precariousness
Precalculus	+	Regional, national networking	Textbook use uncertain in some observed schools
		Perceived quality of text attraction to recruits	Original team moving into other ventures
ChemCom	++	Spinning off variations, translations; new recruits through active dissemination	Original purpose diluted into advanced course for younger students, remedial course for nonacademic students
			Poor collaboration with social sciences and with other chemistry courses

Table 7-7.   **Institutionalization and Continuation of the Eight Reforms and Innovations (continued)**

Project	Likelihood of Continuation	Factors Favoring Continuation	Factors Threatening Continuation
Kids Network	+/-	Appealing formula to recruits	High incidence of discontinuation (cost, outmoded hardware)
		Possibility of integrating new technologies	Deadlines too short; too little computing time
Mimi	+	Likely updating, with better assistance, support	Mandates creating resistance
		New cohorts of teachers more at ease with earlier phases of use	Perception of necessary supplements in content matter

++	Continuation very likely	+/-	Continuation uncertain
+	Continuation likely	-	Continuation unlikely

case studies, none of the projects was threatened outright with extinction, but some looked fragile (Kids Network, and, in some respects, UMC and Mimi), or less likely to spread than innovations (Precalculus) with an assured place in the school curriculum. Others appeared likely to survive, in large part owing to the number of initiatives taken or in progress, but perhaps not at current levels of support (Project 2061, California reform).

Of the remaining projects, the implementation of the NCTM *Standards* is perhaps the hardest to forecast. The *Standards* have established a solid footing at the policy, state and, gradually, district levels, thanks to the initial strong document, the accompanying more operational guides for teaching and assessment, and extensive dissemination. There are, however, national and professional politics involved in standard setting, and no clear protectorate within NCTM to buffer the reform effort against these winds, should they change. Also, more locally, expectations for performance generated from "above" or "outside"— especially if they appear unrealistic—are immediate targets for administrators and teachers insisting on local autonomy and for parents who do not recognize the need for change in the mathematics curriculum.

ChemCom represents an innovation whose basic objective—to reconcile two potentially contradictory objectives, general scientific literacy and a firm place in college preparatory curricula—appeared to be threatened in several ways at sites observed in the case study. Yet, the project continues to gain new adherents and to enjoy support from a powerful professional association. Added to this is its self-financing formula (with royalties plowed back into the project).

The creation of FACETS—Foundations and Challenges to Encourage Technology-Based Science, a set of 24 stand-alone units introducing science on a need-to-know basis—for middle school students and *Chemistry in Context*—a one-semester introductory college course for non-science majors—speaks to the flexibility and ability of ChemCom's leaders to meet various demands for alternative applications of their approach to chemistry courses. Thus, there is a strong likelihood of project continuation, but perhaps not of institutionalization at any given site. Perhaps like the NCTM *Standards*, ChemCom seems destined to take on many guises, for different publics and purposes. Unlike the NCTM *Standards*, it seemed vulnerable at any given site, but strong in its likelihood of further extensions and translations.

Concluding Comments

In many ways, the eight projects we have examined are fundamentally ahead of their time—not in policy terms, where they correspond to the changing conceptions of science and mathematics and new paradigms of teaching and learning, but in structural terms. They are reforms and innovations for which most mainstream institutional settings and instructional practices are not yet ready. Reforms or innovations that come before their time will likely be the targets of explicit conflict or implicit bargaining. Some will be put on hold, blended with ongoing practices and procedures, or watered down. They are unlikely to be routinized. They are fortunate if they can survive in a recognizable form, while profiting from opportunities coming their way.

They are, however, riding on a wave of school science and mathematics reform with greater institutional and political support than projects of this kind have enjoyed up to now. Conceptually, they are in the mainstream. The winds seem to be blowing in their direction, such that, barring a backlash, the next generation of projects should enjoy more support under less precarious conditions. For this to actually transpire, however, such projects will need more advocates and more convincing enactments at local levels. The obvious prerequisite is better prepared and better assisted teachers, ones who are better poised for the leap to the next step on the continuum. The case studies show, however, that a great many other preconditions are needed, not least a committed coalition of constituents in the larger environment.

Coda:
New Paths for Bold Ventures

Michael Huberman

A Sobering Heritage of Reform

Post-mortem studies of large-scale reforms have brought, almost invariably, discouraging news. Large-scale curriculum development projects in mathematics and science, pioneered by the National Science Foundation and the academic community in the 1950s and 1960s, ran wide but shallow for several years. An elaborate process of codevelopment by teachers and university specialists, followed up by study groups, pilot implementation, myriad forms of training and iterative refinements, left few durable traces once external supports were discontinued.[1] Essentially, a wealth of rich, varied, instructionally promising materials lay fallow. Follow-up studies on these and successive innovations in mathematics and science showed few deviations from traditional instruction (Stake and Easley 1978). Similar findings had emerged from the multitude of innovations—some of them large scale—studied by the Rand Corporation (Berman and McLaughlin 1978), and from those followed by the NETWORK Inc. (Crandall et al. 1982).

This is, in many ways, déjà vu—if one goes back to the evaluations of reforms undertaken during the progressive schools movement (Cremin 1961). In Winnetka, Gary, and Denver, for example, structural changes, often combined with changes in curriculum and instructional practice, seldom endured. At the local level, where teachers and students actually interacted, these changes resulted in little instructional impact and were ultimately borne by an increasingly small number of true believers. In his review of large-scale reforms of curriculum and instruction in the late progressive period, Cuban (1984) reports that such practices as pupil-centered pedagogy appeared in no more than a quarter of the classes in districts in which they were systematically introduced (Cuban 1984, p. 135, Elmore 1996). From Cuban's research, in fact, it might be concluded that large-scale reform is an oxymoron, historically speaking.

---

[1] Some exceptions are curricula in high school biology and chemistry. Texts for these courses developed in the 1960s continue, in revised form, to be used in many classrooms today.

More specifically, Cuban found two glaring weaknesses in attempts to instill and generalize large-scale reforms: First, projects developed in one region diffused poorly to another; they were essentially distorted or trivialized. Second, when new, promising instructional enterprises were tracked down to the classroom level, they had almost entirely evaporated.

Studying Bold Ventures in Real Time

What does our study add? Methodologically, it contributes the dimension of following eight reforms and innovations in real time as they unfolded. While there is a rich literature on smaller scale innovations—curriculum or instructional projects affecting many schools but seldom implying institutional and instructional changes of magnitude—there are far fewer studies of large-scale reforms attempting to work their changes simultaneously at a policy level, at state and regional levels, through professional associations, down to districts, and, finally, to the classroom. In four of the eight projects we studied, all these arenas were activated. In three of the remaining four, program developers were after large-scale implementations of experimental curricula. This study has, then, the advantages of scope and immediacy in recording the progress of these projects as they were construed by their protagonists and actually played out. The disadvantage, of course, is a lack of historical detachment. We have tried to account for these reforms and innovations as they were enacted, but could not follow them all the way as they ran their full course.

As noted in earlier chapters, the eight projects shared a common perspective, a common paradigm. All made similar commitments to specific forms of instruction and learning in science and mathematics: links across topics, problem solving or applied approaches to instruction, more "constructivist" views of learners, and concerns for the social implications of science and technology. Some of the projects had other, specific orientations, such as equity (the Urban Mathematics Collaborative—UMC), diversity (Voyage of the Mimi), and generalized scientific literacy (Project 2061).

There is some evidence that these perspectives, as familiar as they seem, are functionally ahead of their time, just as the progressive schools movement was, just as the major reforms in the mathematics and science curricula of the 1960s must have been. One reason for the nonadoption or the distortion of reforms is precisely that the broad band of policymakers, administrators, and teachers are working from a prior paradigm, and can only make sense of what is offered them in terms that are meaningful for them. What the reformers call "distortion" or "trivializing" is, for these actors, simply adaptation to their settings and their logics of use. In several of our cases, as we reported, teachers or trainers were convinced that they were doing constructivist science in their classrooms or with trainees, although there were few observable traces.

Were these eight projects too far ahead of their time, however, we would have seen less activity at the levels of policy and practice; we would also have found less broad-gauged support among key associations and foundations. If there was a time warp, it was narrower than for the previous wave of reforms in this same direction. It was also clear, however, that the main project protagonists were very concerned with strategy. In other words, they understood that they were not in a position to force their reforms into everyday regulation and practice. Project 2061 offered its overarching policy document not as a bible but as a vision-forming guide. The project followed Dewey's (1899) civilized agenda for reform through public discourse and debate (Elmore 1996, p. 11). Note however, that Project 2061 also collaborated loosely with six "experimental" school districts and produced an elaborate set of instructional criteria (*Benchmarks*) for implementing and assessing the approach to science advocated in its policy documents. We see here—and we shall return to this—the multilevel strategies employed by all the major reform projects.

Like Project 2061, the California science education reform trod softly. The state's *Science Framework* was not legislated into practice, nor were external assessments brandished. Rather, a swarm of initiatives was undertaken in what we could characterize as a "particle cloud" strategy of reform (see below). Similarly, the National Council of Teachers of Mathematics (NCTM) project made virtually no attempts to intimidate or threaten its constituencies with its standards, but rather sought to diffuse them gradually; illustrate how they could be applied; and legitimate them as professional mirrors, norms, and challenging models of competent practice.

What we observe here are soft, indirect, and strategic devices for generalizing reforms. They elude conflicts, perhaps deliberately, except at the conceptual level—*among* the NCTM authors, *between* the lead coordinators of the California Science Implementation Network (CSIN) initiative (the network created by California for elementary teacher training in science). The advantages are clear. The disadvantages, as we note below, are that local actors are left with latitude to make of these reforms pretty much what they will. And what they will is often not what the originators had in mind.

New Roles and Strategic Alliances

At the outset, then, the comprehensive reform projects created some form of policy document: a major treatise, as in Project 2061; a mathematics *Standards* document by NCTM; a *Science Framework* in California. What makes these reforms move any farther into the territories of administration and practice?

It is here that the strategies of the eight projects may have broken new ground and, possibly, given a new twist to the process of widespread educational reform. First, some key actors at one level began to work at other levels, in new roles. For example, in Project 2061, central staff and school staff worked together at the "median" level, in the creation of *Benchmarks* for science instruction. Central staff also worked at the local, operational level in coordinating—albeit loosely—with the six experimental school districts. Authors of the NCTM

Standards got involved, with some hesitation, in "selling" the *Standards* at regional and state levels; they became macro-trainers.

Second, projects built coalitions, sometimes unlikely ones. Somewhat anti-establishment teachers in the UMC worked with business and industry. The Precalculus project worked with a private coalition for science and technology education. CSIN put together alliances of California teachers' associations, citizens' groups, teacher education institutes, and research universities. NCTM had perhaps the most elaborate coalition, consisting of professional associations, training and research bodies, federal and state agencies, even private firms. In past reform eras, creating coalitions had been a dissemination task—how to reach the various stakeholders who held the keys to state and local implementation. In the current era, it has become an issue of coordinating a diverse set of agents and agencies already enlisted in the causes espoused by the innovations.

Among the coalitions was a crucial new "player," in many ways a linchpin missing in previous attempts to bridge the federal and local levels: state education agencies. As states have come to play an increasing role in policy, finance, regulation, and accountability, they are now an indispensable part of any broad alliance for reform in science and mathematics education. In several of the projects, state officials were active players.

As the case studies show, coordinating and steering these alliances was no simple task. Up to then, these agents had seldom operated together. Still, by mobilizing the core constituencies that can bring new policies or reform blueprints into the state educational infrastructure, these projects began to accumulate critical mass. In so doing, they also cut down the time—often taking decades—required for reforms to be incarnated in policy documents, frameworks, curriculum guides, and instructional materials. An effective tactic appears to have been to focus on common, superordinate goals, and to play down the inevitable conflicts of interest among the parties. For example, state-level officials were interested in standards for accountability and performance; professional associations wanted challenging levels of professional competence; some citizens' groups were after levels of learning that all students, including the most disadvantaged, could achieve. Thus, all could be united behind standards-based reform.

Systems and Particle Clouds

The California science education reform was depicted in detail in the case study of Atkin et al. (1996c). It can be viewed in many ways. We have chosen to describe it here as a particle cloud strategy of reform. It was neither top-down nor bottom-up, but was rather both at once. Groups of teachers helped frame and revise the initiative, but there would have been no initiative to work with at all had there not been a central vision, an orienting framework, apparently audacious state-level leadership, and a coherent strategy.

The same can be said of the NCTM standards project, which was essentially a peer reform: that is, a document drafted by acknowledged experts and accepted peers for the larger community of mathematics educators and teachers. That the first *Standards* document was meant to be professionally binding, and to be followed by others that addressed more closely the realities of everyday mathematics instruction, gave it local legitimacy as well as professional adherence. It was not, however, a grassroots reform or a federal-state initiative. When the *Standards* document intersected with district or state accountability measures, however, it could easily be turned into a top-down assessment mandate.

The California reform had other intriguing properties. It launched a thousand small ships: The *Framework* gave rise to small initiatives in teacher education programs in universities; in the creation of networks and pilot projects; in the assignment of teachers as mentors, consultants, and curriculum developers; in study groups and in-service offerings; in prototype materials for use in schools; in various kinds and grain sizes of publications. Unlike many reforms in this study and elsewhere, it was hard for teachers or principals *not* to have been affected by the initiative in some, albeit slight, way. Even training for administrators was provided for. In other words, actors in science and mathematics education throughout the state were *surrounded* by the reform. Thus the particle cloud metaphor.

The California strategy is systemic in that it affects all sectors of activity. But it is peculiarly systemic. For example, it has too many points of activity to be "engineered." It can be rationally designed, intelligently launched, and—to some extent—deliberately reoriented or reactivated if it goes astray, is overly misshapen, or dwindles. But only up to a point. Too many ships have been launched into many different weather systems. It has deliberately overreached. Fundamentally, the initiators must let go, much as in the administration of the Internet. The enterprise is not rudderless, but no one "controls" it. Any control, moreover, has to be improvisational. This may represent, then, a new model of educational reform, one which accommodates the unknown, unpredictable, and Murphy's law-like nature of the contexts within which educational policies and practices actually operate. What is hard to determine is how precarious this model might be.

The other large-scale reform projects are equally systemic, but in the more conventional sense of having vertical ties between local practices and general policies at the state, regional or national level. This linkage was not always strong, as the case studies show. In effect, both Project 2061 and NCTM standards are faint presences in the schools studied. But the links, even if weak, did lend coherence to the project as a whole. In the curriculum development projects, by contrast, there was far more dilution, distortion, or outright resistance when the project moved from developer to publisher, and then to widespread classroom implementation. In the Voyage of the Mimi, for example, a multimillion dollar investment in inventive computer software, print materials, video series, and an accompanying set of guides for teachers and pupils was gradually winnowed down to instructional set pieces that exploited fewer than 5 percent of the original materials. There was a similar dilution in the Kids Network, and some curious spin-offs of ChemCom in participating schools. This applications-rich program for non-science majors was used in some schools as a basic chemistry curriculum for advanced students at a lower grade level; in others, it was used for students deemed unable to handle the traditional 11th grade chemistry course. This sort of adaptation made the program more generally popular, but obscured its original purpose. And the Precalculus textbook may have been so faithful to the vision of its originators that only they could use it in the ways it was intended.

In other words, these smaller scale, stand-alone projects may well have suffered the traditional fate of innovations—dilution, withering, distortion, trivializing, reversion to traditional practice—precisely because they were not hitched to larger ventures at the state or national levels. To be sure, questions of local readiness, institutional and individual capacity, internal and external support, willingness to endure uncertainties and exhaustion until adequate levels of mastery are established all characterize the local enactments of the eight cases (Huberman and Miles 1984). In the reform projects, including UMC, however, the distortion/trivializing process appeared to be more in flux and more negotiable. For the smaller projects, it seemed as if some might well suffer the fate of 1960s efforts.

How Much Reform, Anyway?

The challenge of the large-scale reform projects in our sample is that of achieving critical mass without the means of servicing it, or even really managing it beyond the phase of initiation. In particular, the squeeze comes at the local levels, in the schoolhouse. Reformers who have comfortably written vision statements or, less comfortably, hammered out agreements on wording and demands are not in the place of those who are meant to enact changes in their instructional environments—changes that, by definition, no one has debugged. The challenge is to provide the resources, organizational conditions, and technical support that make teachers unwilling to go back to conventional instruction in mathematics and science.

It is impossible to estimate how many classrooms actually engaged with these reforms during the period of our work. How many mathematics teachers in the cities hosting UMC actually participated in the collaboratives, and to what extent? How many California teachers actually modified their instructional environments as a result of the state's reform, and were these trivial or momentous changes in practices? How many mathematics teachers revised their syllabus or even their calendar as a result of the NCTM standards? It is hard to say. As noted earlier, a historical ballpark estimate, at the peak of reform periods, has roughly 25 percent of teachers involved out of a total population of potential participants.

From estimates in the case studies, the California reform affected the largest proportion of actors, but only in one state and not necessarily at the level of actual instruction. There is a tendency among reformers—and perhaps among case researchers—to amplify the magnitude of the change taking place and under consideration, often because that is where they are shining their flashlight. Still, looking closely at the case studies of the NCTM standards, the California reform, and UMC, it would seem that a large number of actors—including both teachers and teacher educators—were actually experimenting with nontrivial materials and practices that were novel to them. As modest as the numbers may be in absolute terms, this is a highly significant development in mathematics and science education.

Beyond that assertion, one must take the long view. The issue of measurable outcomes is one that educational historians will have to resolve, and they will be at pains to do so. These reforms, much like the paradigm that forms their context, will join with others, large and small, now emerging or still to emerge, to create whatever instructional impacts can be attributed to them. If, moreover, the tide flows back to an earlier paradigm—as with the current revival of "the basics" running in parallel to the constructivist perspective—we will not know how to evaluate the (presumably diminished) effects of contemporary reforms.

The Future of the Present

We would be hard put to assess either the rate of growth/decline or the longevity of these eight projects. In chapter 7, we risk some estimates, however. There is evidence that the California reform is under pressure and has been cut back. If, however, the particle cloud theory is correct, we would expect it to scale back— and then scale back up when the environment changes. The reform appears to have largely become the fishbowl in which the majority of science educators in the state now swim, which is a considerable achievement. Similarly, NCTM appears to be contending with intrapolitical and professional issues, yet continues on course with the publication of new standards documents. These documents have increasingly become reference points for the design of curricula and assessments. As such, they are rooted—until they are supplanted or they build in their own revisions. Project 2061 and UMC appear more fragile, given their heavy dependence on foundation support. For the most part, the curriculum development projects (Mimi, Kids Network, ChemCom) will live or die by their upgrades, sales, and service—although ChemCom enjoys continuing support from the American Chemical Society. Finally, the Precalculus project will likely thrive in its own institutional surround and continue to disseminate its successful textbook, but only venture out if its core staff feels the call to take others along on their pilgrimage.

Adelman, N., T. Corcoran, E. Hawkins, P. Shields, and A. Zucker. 1995. *Evaluation of the National Science Foundation's Statewide Systemic Initiatives (SSI) Program: Second year report.* Contractor report submitted under NSF contract # SED-9255371 Menlo Park, CA: SRI International.

Alexander, P. A. 1996. Of squalls and fathoms: Navigating the seas of educational innovation. *Educational Researcher* 25(4): 31-39.

American Association for the Advancement of Science (AAAS). 1989. *Science for all Americans. A Project 2061 report on literacy goals in science, mathematics, and technology.* Washington, DC.

———. 1993. *Benchmarks for science literacy.* New York: Oxford University Press.

———. 1995. *Project 2061: Science literacy for a changing future, A decade of reform.* Washington, DC.

American Chemical Society. 1988a. *ChemCom: Chemistry in the community.* Teacher's Guide. 1st ed. Dubuque, IA: Kendall/Hunt.

———. 1988b. *ChemCom: Chemistry in the community.* 1st ed. Dubuque, IA: Kendall/Hunt.

American Federation of Teachers (AFT). 1995. *Setting strong standards: AFT's criteria for judging the quality and usefulness of student achievement standards.* Washington, DC.

Apple, M. 1992. Do the standards go far enough? Power, policy, and practice in mathematics education. *Journal for Research in Mathematics Education* (5): 412-31.

Atkin, J. 1994. Developing world-class education standards: Some conceptual and political dilemmas. In *The future of education: Perspectives on national standards in America,* ed. N. Cobb, 61-84. New York: College Entrance Examination Board.

Atkin, J. M., and A. Atkin. 1989. *Improving science education through local alliances.* New York: Carnegie Corporation.

Atkin, J. M., P. Black, E. D. Britton, and S. A. Raizen 1996a. A global revolution in science, mathematics, and technology education. *Education Week* April 10.

Atkin, J. M., J. A. Bianchini, and N. I. Holthuis. 1996b. The different worlds of Project 2061. In *Bold ventures, volume 2: Case studies of U.S. innovations in science education,* eds. S. A. Raizen and E. D. Britton, 131-246. Boston: Kluwer Academic Press.

Atkin, J. M., J. V. Helms, G. L. Rosiek, and S. A. Siner. 1996c. Building on strength: Changing science teaching in California public schools. In *Bold ventures, volume 2: Case studies of U.S. innovations in science education,* eds. S. A. Raizen and E. D. Britton, 13-130. Boston: Kluwer Academic Press.

Ball, D. 1988. *Research on teaching mathematics: Making subject matter knowledge part of the equation.* East Lansing, MI: National Center for Research on Teacher Learning, Michigan State University.

_____. 1990. *Implementing the NCTM Standards. Hopes and hurdles.* East Lansing, MI: National Center for Research on Teacher Learning, Michigan State University.

Barrett, G. B., K. G. Bartkovich, H. L. Compton, S. Davis, D. Doyle, J. A. Goebel, L. D. Gould, J. L. Graves, J. A. Lutz, and D. J. Teague. 1992. *Contemporary precalculus through applications: Functions, data analysis and matrices.* Rev. ed. Dedham, MA: Janson Publications.

Bereiter, C. 1994. Constructivism, socioculturalism, and Popper's world 3. *Educational Researcher* 23(7): 21-23.

Berlak, H., F. Newmann, E. Adams, D. Archbald, T. Burgess, J. Raven, and T. Romberg. 1992. *Toward a new science of educational testing and assessment.* Albany, NY: State University of New York Press.

Berman, P., and M. McLaughlin. 1978. *Federal programs supporting educational change. Vol. VIII: Implementing and sustaining innovations.* Santa Monica, CA: Rand Corporation.

Black, P. 1994. Performance assessment and accountability: The experience in England and Wales. *Educational Evaluation and Policy Analysis* 16(2): 191-203.

Black, P., and J. M. Atkin, eds. 1996. *Changing the subject: Innovations in science, mathematics and technology education.* New York: Routledge.

Blair, L. H., P. Brounstein, H. Hatry, and E. Morley. 1990. *Guidelines for school business partnerships in science and mathematics.* Washington, DC: The Urban Institute.

Blank, R., and D. Gruebel. 1993. *State indicators of science and mathematics education 1993.* Washington, DC: Council of Chief State School Officers.

Blank, R., and E. M. Pechman. 1995. *State curriculum frameworks in mathematics and science: How are they changing across the states?* Washington, DC: Council of Chief State School Officers.

Bracey, G. W. 1996. International comparisons and the condition of American education. *Educational Researcher* 25(1): 5-11.

Britton, E. 1993a. California's systemic improvement of science education. In *Science and mathematics education in the United States: Eight innovations*, 19-59. Washington, DC: Organisation for Economic Co–operation and Development.

_____. 1993b. Kids Network. In *Science and mathematics education in the United States: Eight innovations*, 75–94. Washington, DC: Organisation for Economic Co–operation and Development.

Brophy, J., and J. Alleman. 1991. Activities as instructional tools: A framework for analysis and evaluation. *Educational Researcher* 20(4): 9-23.

Bussis, A. M., E. A. Chittenden, and M. Amarel. 1976. *Beyond surface curriculum.* Boulder: Westview Press.

California Department of Education. 1985. *Mathematics Framework for California public schools: Kindergarten through grade twelve.* Sacramento, CA.

_____ . 1990. *Science framework for California public schools kindergarten through grade twelve.* Sacramento, CA.

_____ . 1995. *A sampler of science assessment.* Sacramento, CA.

Callahan, R. 1962. *Education and the cult of efficiency.* Chicago: University of Chicago Press.

Carnegie Forum on Education and the Economy. 1986. *A nation prepared: Teachers for the 21st century.* A report of the Task Force on Teaching as a Profession. Washington, DC.

Catlin, S. 1986. New kids in class: Minorities pace growth in public schools. Analysis of State Government and Politics. *California Journal* XVII(4): 189-92.

Char, C., and J. Hawkins. 1987. Charting the course: Involving teachers in the formative research and design of the Voyage of the Mimi. In *Mirrors of minds: Patterns of experience in educational computing,* eds. R. D. Pea and K. Sheingold, 211-22. Norwood, NJ: Ablex Publishing Corporation.

Char, C., J. Hawkins, J. Wooten, K. Sheingold, and T. Roberts. 1983. *Voyage of the Mimi: Classroom case studies of software, video, and print materials. Phase 1.* New York: Bank Street College of Education.

Chubin, D. 1995. Hot House? Research on systemic reform: An activity proposed for NISE. University of Wisconsin at Madison: National Institute for Science Education.

Cobb, P. 1994. Where is the mind? Constructivist and sociocultural perspectives on mathematical development. *Educational Researcher* 23(7): 13-20.

Cohen, D. 1990. A revolution in one classroom: The case of Mrs. Oublier. *Educational Evaluation and Policy Analysis* 12: 327-45.

_____ . 1995. What is the system in systemic reform? *Educational Researcher* 24(9): 11-17.

Cohen, D., D. L. Ball, P. Peterson, N. J. Wiemers, S. M. Wilson, L. Darling-Hammond, and G. Skyes. 1990. *Educational Evaluation and Policy Analysis* 12(3).

Coley, R., and M. Goertz. 1990. *State educational standards in the fifty states: 1990.* ETS RR 9S15. Princeton: Educational Testing Service.

Commission on the Skills of the American Workforce. 1990. *America's choice: High skills or low wages!* Rochester, NY: National Center on Education and the Economy.

Crandall, D. 1982. *People, policies and practice: Examining the chain of school improvement.* Andover, MA: The NETWORK, Inc.

Cremin, L. 1961. *The transformation of the American school.* New York: Knopf.

_____ . 1964. *The transformation of the school: Progressivism in American education 1876-1957.* New York: Vintage Books, Random House.

Cronbach, L. 1977. *Aptitudes and instructional methods: A handbook for research on interactions.* San Francisco: Jossey-Bass.

Crosswhite, F. J. 1990. National standards: A new dimension in professional leadership. *School science and mathematics* 454-66.

Crosswhite, F. J., J. A. Dossey, and S. M. Frye. 1989. NCTM standards for school mathematics: Visions for implementation. *Arithmetic Teacher* (3): 55-60.

Cuban, L. 1984. *How teachers teach: Constancy and change in American classrooms, 1890-1980*. New York: Longman.

Darling-Hammond, L. 1990. Achieving our goals: Superficial or structural reforms. *Phi Delta Kappan* 72(4): 286-95.

Dewey, J. 1899. *The school and society*. Chicago: University of Chicago Press.

Donmoyer, R. 1995. The rhetoric and reality of systemic reform: A critique of the proposed national science education standards. *COGNOSOS—The National Center for Science Teaching and Learning Research Quarterly* 4(1).

Driver, R. 1988. Theory into practice II: A constructivist approach to curriculum. In *Development and dilemmas in science education*, ed. P. Fenshaw, 133-49. London: Falmer Press.

Driver, R., H. Asoko, J. Leach, E. Mortimer, and P. Scott. 1994. Constructing scientific knowledge in the classroom. *Educational Researcher* 23(7): 5-12.

Educational Evaluation & Policy Analysis. 1990. Entire thematic issue (devoted to the findings of the California study of elementary mathematics). 12(3) Fall.

Eisenhart, M., E. Finkel, and S. F. Marion. 1996. Creating the conditions for scientific literacy: A re–examination. *American Educational Research Journal* 33(2): 261-65.

Eisner, E. 1992. Curriculum ideologies. In *Handbook of research on curriculum: A project of the American Educational Research Association*, ed. P. W. Jackson, 302-24. New York: Macmillan.

Elmore, R. 1996. Getting to scale with good educational practice. *Harvard Educational Review* 66(1): 1-26.

Federal Coordinating Council for Science, Engineering, and Technology (FCCSET). 1992. *By the year 2000: First in the world*. A report from the Committee on Education and Human Resources. Washington, DC.

_____. 1993. *Pathways to excellence: A federal strategy for science, mathematics, engineering, and technology education*. A report from the Committee on Education and Human Resources. Washington, DC.

Ferrini-Mundy, J., and M. A. Clements. 1994. Recognizing and recording reforms in mathematics—New questions, many answers. *Mathematics Teacher* (88): 380-89.

Fuhrman, S., and D. Massell. 1992. *Issues and strategies in systemic reform*. CPRE Research Report Series RR-025. New Brunswick, NJ: Consortium for Policy Research in Education.

Fullan, M. 1991. *The new meaning of educational change*. New York: Teachers College Press.

_____. 1993. *Change forces: Probing the depths of educational reform*. Bristol, PA: Falmer Press.

_____. 1994. Coordinating top-down and bottom-up strategies for educational reform. In *The governance of curriculum*, eds. R. Elmore and S. Fuhrman, 186-202. Alexandria, VA: Association for Supervision and Curriculum Development.

Grossman, P., S. Wilson, and L. Shulman. 1989. Teachers of substance: Subject matter knowledge for teaching. In *Knowledge base for the beginning teacher*, ed. M. Reynolds, 23-36. New York: Pergamon.

Guba, E. 1968. The process of educational innovation. In *Educational change*, ed. R. Goulet, 136-53. New York: Citation Press.

Hall, G. 1977. The study of individual teacher and professor concerns about innovations. *Journal of Teacher Education* 28(11): 22-23.

Hall, G., and S. Loucks-Horsley. 1978. A developmental model for determining whether the treatment is actually implemented. *American Educational Research Journal* 14(3): 263-76.

Hall, G., S. Loucks–Horsley, S. Rutherford, and B. Newlove. 1975. Levels of use of the innovation: A framework for analyzing innovation implementation. *Journal of Teacher Education* 16(1): 52-56.

Havelock, R. 1969. *Planning for innovation through the creation and dissemination of knowledge*. Ann Arbor, MI: ISR, University of Michigan.

Hesse-Bieber, S., T. S. Kinder, P. R. Dupuis, A. Dupuis, and E. Tornabene. 1993. HyperRESEARCH [computer software]. Randolph, MA: ResearchWare.

Honig, B. 1994. How can Horace be helped? *Phi Delta Kappan* (75): 790-96.

Huberman, M. 1995. Professional careers and professional development: Some intersections. In *Professional development in education*, eds. T. Guskey and M. Huberman, 193-224. New York: Teachers College Press.

Huberman, M., and M. Miles. 1984. *Innovation up close: How school improvement works*. New York: Plenum.

Humphrey, D., and P. M. Shields. 1996. *Study of the Dwight D. Eisenhower Mathematics and Science Education State Curriculum Frameworks projects and regional consortia program: A review of mathematics and science curriculum frameworks*. Contractor report submitted under contract #EA9306 to U.S. Department of Education, Office of the Under Secretary of Education, Planning and Evaluation Service. Menlo Park, CA: SRI International.

Hutchins, C. L. 1996. *Systemic thinking: Solving complex problems*. Aurora, CO: Professional Development Systems.

Hutchinson, J., and M. Huberman. 1993. *Knowledge dissemination and use in science and mathematics education: A literature review*. Washington, DC: National Science Foundation.

International Association for the Evaluation of Educational Achievement (IEA). 1988. *Science achievement in seventeen countries: A preliminary report*. Elmsford, NY: Pergamon Press.

Jackson, P. W., ed. 1992. *Handbook of research on curriculum: A project of the American Educational Research Association*. New York: Macmillan.

Karlan, J. W., M. Huberman, and S. H. Middlebrooks. 1996. The challenges of bringing the Kids Network to the classroom. In *Bold ventures, volume 2: Case studies of U.S. innovations in science education*, eds. S. A. Raizen and E. D. Britton, 247-396. Boston: Kluwer Academic Press.

Kaser, J., and S. Loucks-Horsley. 1995. *Profiling systemic programs: A summary report*. Andover, MA: The NETWORK, Inc.

Kilpatrick, J., and G. M. A. Stanic. 1995. Paths to the present. In *Seventy-five years of progress: Prospects for school mathematics*, ed. I. M. Carl, 3-17. Reston, VA: National Council of Teachers of Mathematics.

Kilpatrick, J., L. Hancock, D. S. Mewborn, and L. Stallings. 1996. Teaching and learning cross-country mathematics: A story of innovation in precalculus. In *Bold ventures, volume 3: Case studies of U.S. innovations in mathematics education*, eds. S. A. Raizen and E. D. Britton, 133-245. Boston: Kluwer Academic Press.

Knapp, M. S. 1995. *Education policy and the improvement of teaching: Two accounts of the early implementation of the California Mathematics Framework*. Presented at the annual meeting of the American Educational Research Association, San Francisco.

Lapointe, A. E., J. M. Askew, and N. A. Mead. 1992. *Learning science*. Princeton: Educational Testing Service.

Lapointe, A. E., N. A. Mead, and G. W. Phillips. 1989. *A world of differences*. Princeton: Educational Testing Service.

Lindquist, M. M. 1994a. *Lessons learned?* Paper presented at the Adult Mathematical Literacy Conference, Arlington, VA.

_____. 1994b. Linking yesterday to tomorrow. *Teaching Children Mathematics*, 1: 53-58.

Louis, K. S., and S. Rosenblum. 1981. *Linking R&D with schools: A program and its implications for dissemination and school improvement policy*. Washington, DC: U.S. Office of Education.

Lund, L., and C. Wild. 1993. *Ten years after a nation at risk*. New York: The Conference Board.

Lynch, M., and E. Britton. 1993a. Chemistry in the community (ChemCom). In *Science and mathematics education in the United States: Eight innovations*, 59-74. Washington, DC: Organisation for Economic Co–operation and Development.

_____. 1993b. Project 2061. In *Science and mathematics education in the United States: Eight innovations*, 95-116. Washington, DC: Organisation for Economic Co–operation and Development.

Marsh, D. D., and A. R. Odden. 1991. Implementation of the California Mathematics and Science Frameworks. In *Educational policy implementation*, ed. A. R. Odden, 219-40. Albany, NY: State University of New York Press.

Martin, L. 1985a. *Interim report, mathematics, science and technology teacher project, Phase I: Evaluation of training*. New York: Bank Street College of Education.

———. 1985b. *Progress report, Bank Street College of Education, mathematics, science and technology teacher project*. New York: Bank Street College of Education.

Martin, L., J. Hawkins, S. Y. Gibbon, and R. McCarthy. 1988. Integrating information technologies into instruction: The Voyage of the Mimi. In *AETS Yearbook: Information technology and science education*, ed. J. D. Ellis, 173-86. Columbus, OH: ERIC Clearinghouse for Science, Mathematics and Environmental Education, The Ohio State University.

Massell, D. 1994. National curriculum content standards: The challenges for subject matter associations. In *The future of education*, ed. N. Cobb, 239-57. New York: College Entrance Examination Board.

Mathematical Sciences Education Board. 1989. *Everybody counts*. Washington, DC: National Academy Press.

McLaughlin, M. W. 1990. The Rand change agent study revisited: Macro perspectives and micro realities. *Educational Researcher* 19(9): 11-16.

McLeod, D. B., R. E. Stake, B. P. Schappelle, M. Mellissinos, and M. J. Gierl. 1996. Setting the standards: NCTM's role in the reform of mathematics education. In *Bold ventures, volume 3: Case studies of U.S. innovations in mathematics education*, eds. S. A. Raizen and E. D. Britton, 13-132. Boston: Kluwer Academic Press.

Middlebrooks, S. H., M. Huberman, and J. W. Karlan. 1996. Science, technology, and story: Implementing the Voyage of the Mimi. In *Bold ventures, volume 2: Case studies of U.S. innovations in science education*, eds. S. A. Raizen and E. D. Britton, 395-518. Boston: Kluwer Academic Press.

Miles, M., and M. Huberman. 1994. *Qualitative data analysis: A sourcebook of new methods*. 2nd edition. Thousand Oaks, CA: Sage.

Mullis, I., J. Dossey, M. Foertsch, L. Jones, and C. Gentile. 1991. *Trends in academic progress: Achievement of U.S. students in science 1969-1970 to 1990; mathematics, 1973-1990; reading, 1971-1990;* and *writing, 1984-1990*. Princeton: Educational Testing Service.

Myers, M. 1994. Problems and issues facing the National Standards Project in English. In *The future of education*, ed. N. Cobb, 259-76. New York: College Entrance Examination Board.

National Academy of Sciences. 1982. *Science and mathematics in the schools: Report of a convocation*. Washington, DC: National Academy Press.

National Center for Education Statistics (NCES). 1992. *Digest of education statistics: 1992*. A report of the Office of Educational Research and Improvement, U.S. Department of Education. NCES 92–097. Washington, DC: U.S. Government Printing Office.

National Commission on Excellence in Education. 1983. *A nation at risk: The imperative for educational reform*. Washington, DC: U.S. Government Printing Office.

National Council of Teachers of Mathematics (NCTM). 1980. *An agenda for action: Recommendations for school mathematics of the 1980s*. Reston, VA.

_____ . 1989. *Curriculum and evaluation standards for school mathematics*. Reston, VA.

_____ . 1991. *Professional standards for teaching mathematics*. Reston, VA.

_____ . 1995. *Assessment standards for school mathematics*. Reston, VA.

National Governors' Association. 1990. *National education goals*. Washington, DC.

National Research Council (NRC). 1996. *National science education standards*. Washington, DC.

National Science Board, Commission on Pre-college Education in Mathematics, Science and Technology. 1983. *Educating Americans for the 21st century*. Washington, DC: National Science Foundation.

National Science Foundation (NSF). 1990a. *Statewide systemic initiatives in science, mathematics, and engineering education: Program solicitation*. NSF 90-47. Washington, DC.

_____ . 1990b. *Private sector partnerships to improve science and mathematics education: Program solicitation*. NSF 90-18. Washington, DC.

_____ . 1994a. *Local systemic change through teacher enhancement grades K-8: Program solicitation and guidelines*. NSF 94-73. Washington, DC.

_____ . 1994b. *Women, minorities, and persons with disabilities in science and engineering*. Arlington, VA.

_____ . 1995. The systemic reform "report card." Draft. Arlington, VA: Division of Educational System Reform Programs.

_____ . 1996. *Indicators of science and mathematics education,* ed. L. Suter. NSF 96-52. Arlington, VA: Division of Research, Evaluation, and Communication, Directorate for Education and Human Resources.

Nelson, B. S. 1994. Mathematics and community. In *Reforming mathematics education in America's cities: The Urban Mathematics Collaborative project*, eds. N. L. Webb and T. A. Romberg, 8-23. New York: Teachers College Press.

O'Day, J., and M. Smith. 1993. Systemic reform and educational opportunity. In *Designing coherent education policy: Improving the system*, ed. S. H. Fuhrman. San Francisco: Jossey-Bass.

Ogborn, J. 1995. Recovering reality. *Studies in Science Education* 25: 3-38.

Organisation for Economic Co–operation and Development (OECD). 1993. *Science and mathematics education in the United States: Eight innovations.* Washington, DC.

Phillips, D. C. 1995. The good, the bad, and the ugly: The many faces of constructivism. *Educational Researcher* 24(7): 5-12.

Pintrich, P. R., R. W. Marx, and R. A. Boyle. 1993. Beyond cold conceptual change: The role of motivational beliefs and classroom contextual factors in the process of conceptual change. *Review of Educational Research* 63(2): 167-99.

Porter, A. 1989. External standards and good teaching: The pros and cons of telling teachers what to do. *Educational Evaluation and Policy Analysis* 11(4): 343-56.

Porter, A., R. Floden, D. Freeman, W. Schmidt, and J. Schwille. 1988. Content determinants in elementary school mathematics. In *Perspectives on research on effective mathematics teaching*, eds. D. Grouws and T. Cooney, 96-113. Hillsdale, NJ: L. Erlbaum.

Porter, A., M. W. Kirst, E. J. Osthoff, J. L. Smithson, and S. A. Schneider. 1993. *Reform up close: An analysis of high school mathematics and science classrooms.* Final report. Madison, WI: Wisconsin Center for Education Research, University of Wisconsin.

Prawat, R. 1989. Promoting access to knowledge strategy and disposition in students: A research synthesis. *Review of Educational Research* 59(1): 1-41.

Price, J. 1994. Reform is a journey, not a destination. *NCTM News Bulletin* 31(4): 3.

Ragin, C. 1987. *The comparative method. Moving beyond qualitative and quantitative strategies.* Berkeley: University of California Press.

Raizen, S. A. 1991. The reform of science education in the U.S.A.: Déjà vu or de novo?. *Studies in Science Education* 19: 1-41.

_____ . 1996. *Standards for science education.* University of Wisconsin-Madison: National Institute for Science Education.

Raizen, S. A., P. Sellwood, R. D. Todd, and M. Vickers. 1995. *Technology education in the classroom: Understanding the designed world.* San Francisco: Jossey-Bass.

Raizen, S. A., and E. D. Britton, eds. 1996a. *Bold ventures, volume 2: Case studies of U.S. innovations in science education.* Boston: Kluwer Academic Press.

_____ . 1996b. *Bold ventures, volume 3: Case studies of U.S. innovations in mathematics education.* Boston: Kluwer Academic Press.

Ravitch, D. 1995. *National standards in American education: A citizen's guide.* Washington, DC: The Brookings Institution.

Reynolds, W. W., Jr., I. H. Gawley, Jr., and F. T. Pregger. 1993. *Impact study of 10 NSF-supported pre-college instructional materials projects.* Final report to the National Science Foundation under grant ESI 9253093. Haddonfield, NJ: Reynolds & Associates, Inc., Communications.

Rogers, E. 1962. *Diffusion of innovations*. New York: Macmillan.

_____ . 1983. *Diffusion of innovations*. 2nd edition. New York: Free Press.

_____ . 1988. The intellectual foundation and history of the agricultural extension model. *Knowledge* 9(4): 492-510.

Romberg, T. A. 1984. *Curricular reform in school mathematics: Past difficulties, future possibilities*. Paper prepared for the Fifth International Congress on Mathematical Education, Adelaide, South Australia.

Romberg, T. A., and A. Pitman. 1994. A strategy for social change. In *Reforming mathematics education in America's cities: The Urban Mathematics Collaborative project*, eds. N. L. Webb and T. A. Romberg, 48-66. New York: Teachers College Press.

Romberg, T. A., and N. L. Webb. 1993a. The role of the National Council of Teachers of Mathematics in the current reform movement in school mathematics in the United States of America. In *Science and mathematics education in the United States: Eight innovations*, 143-82. Washington, DC: Organisation for Economic Co-operation and Development.

_____ . 1993b. The Urban Mathematics Collaborative project: C^2ME as a case study on teacher professionalism. In *Science and mathematics education in the United States: Eight innovations*, 183-206. Washington, DC: Organisation for Economic Co-operation and Development.

Rowe, M. B., J. E. Montgomery, M. J. Midling, and T. M. Keating. 1996. ChemCom's evolution: Development, spread, and adaptation. In *Bold ventures, volume 2: Case studies of U.S. innovations in science education*, eds. S. A. Raizen and E. D. Britton, 519-84. Boston: Kluwer Academic Press.

Sarason, S. 1971. *The culture of the school and the problem of change*. Boston: Allyn and Bacon.

Senge, P. 1990. *The fifth discipline*. New York: Doubleday.

Shavelson, R. J., and G. P. Baxter. 1991. Performance assessment in science. In *Applied Measurement in Education* 4(4): 347-62.

Smith, M., and J. O'Day. 1991. Systemic school reform. In *Politics of curriculum and testing, the 1990 yearbook of the politics of education*, eds. S. Fuhrman and B. Malen, 233-67. London: The Falmer Press.

Snow, R., and E. Mandinach. 1991. *Integrating assessment and instruction: A research and development agenda*. RR-91-8. Princeton: Educational Testing Service.

Solomon, J. 1994. The rise and fall of constructivism. In *Studies in Science Education* 23: 1-19. London: Studies in Education Ltd.

Sparkes, G., and S. Loucks-Horsley. 1990. Models of staff development. In *Handbook of research on teacher education*, ed. W. Houston, 234-50. New York: Macmillan.

Stake, R. 1995. The invalidity of standardized testing for measuring mathematics achievement. In *Reform in school mathematics and authentic assessment*, ed. T. A. Romberg. Albany: State University of New York Press.

Stake, R. E., and J. A. Easley. 1978. *Case studies in science education.* Urbana, IL: Center for Instructional Research and Curriculum Evaluation, University of Illinois.

Stallings, J. 1989. *School achievement effects and staff development. What are some critical factors?* Paper presented at American Educational Research Association, San Francisco.

Stanic, G. M. A., and J. Kilpatrick. 1992. Mathematics curriculum reform in the United States: A historical perspective. *International Journal of Educational Research* 17: 407-17.

Stodolsky, S. 1988. *The subject matters: Classroom activity in mathematics and social studies.* Chicago: University of Chicago.

Sussman, A., ed. 1993. *Science education partnerships.* San Francisco: University of California.

Task Force on Education for Economic Growth. 1983. *Action for excellence: A comprehensive plan to improve our nation's schools.* Denver: Education Commission of the States.

Technical Education Research Centers (TERC). 1987. *National Geographic Society Kids Network project: Annual report October 1, 1986-September 30, 1987.* Cambridge, MA.

_____ . 1989. *National Geographic Society Kids Network project: Annual report October 1, 1988-September 30, 1989.* Cambridge, MA.

Triangle Coalition for Science and Technology Education. 1991. *A guide for building an alliance for science, mathematics, and technology education.* College Park, MD.

_____ . 1994. *Annual report 1993.* College Park, MD.

Twentieth Century Fund Task Force. 1983. *Report of the Twentieth Century Fund Task Force on federal elementary and secondary education policy.* New York.

von Glasersfeld, E. 1995. *Radical constuctivism: A way of knowing and learning.* London: Falmer Press.

Webb, N. L. 1993a. Mathematics accessible through technology: The voyage of the Mimi as an interdisciplinary and technologically-based program. In *Science and mathematics education in the United States: Eight innovations,* 207-28. Washington, DC: Organisation for Economic Co-operation and Development.

_____ . 1993b. State of California: Restructuring of mathematics education. In *Science and mathematics education in the United States: Eight innovations,* 117-43. Washington, DC: Organisation for Economic Co-operation and Development.

Webb, N. L., and T. A. Romberg, eds. 1994. *Reforming mathematics education in America's cities: The Urban Mathematics Collaborative project.* New York: Teachers College Press.

Webb, N. L., D. J. Heck, and W. F. Tate. 1996. The Urban Mathematics Collaborative project: A study of teacher, community, and reform. In *Bold ventures, volume 3: Case studies of U.S. innovations in mathematics education*, eds. S. A. Raizen and E. D. Britton, 245-360. Boston: Kluwer Academic Press.

Webb, N. L., H. Schoen, and S. D. Whitehurst. 1993. *Dissemination of nine precollege mathematics instructional materials projects funded by the National Science Foundation, 1981-91*. Final report to the National Science Foundation under grant MDR 9252727. Madison, WI: Wisconsin Center for Education Research, University of Wisconsin.

Weick, K. 1979. Educational organizations as loosely coupled systems. *Administrative Science Quarterly* 21: 1-19.

_____ . 1984. Small wins: Redefining the scale of social problems. *American Psychologist* 39(1): 40-49.

Weiss, I. R. 1978. *Report of the 1977 national survey of science, mathematics and social studies education*. Research Triangle Park, NC: Research Triangle Institute.

_____ . 1987. *Report of the 1985-86 national survey of science and mathematics education*. Research Triangle Park, NC: Research Triangle Institute.

_____ . 1994. *A profile of science and mathematics education in the United States, 1993*. Draft report. Chapel Hill, NC: Horizon Research, Inc.

Weiss, I. R., M. C. Matti, and P. Smith. 1994. *Report of the 1993 national survey of science and mathematics education*. Chapel Hill, NC: Horizon Research, Inc.

Welch, W. W. 1979. Twenty years of science curriculum development: A look back. In *Review of research in education 7*, ed. D. C. Berliner, 282-306. Washington, DC: American Educational Research Association.

Werner, H., and B. Kaplan. 1963. *Symbol formation*. New York: Wiley.

Yin, R. 1978. *Changing urban bureaucracies: How new practices become routinized*. Santa Monica, CA: Rand Corporation.

Yin, R., S. K. Quick, P. M. Bateman, and E. L. Marks. 1978. *The routinization of innovations*. Santa Monica, CA: Rand Corporation.

Appendix:
Case Studies in Other Countries

As shown in the table below, the United States and 12 other member countries in the Organisation for Economic Co-operation and Development studied 23 innovations in science, mathematics, or technology education. While U.S. researchers looked at eight innovations, countries generally selected a single innovation. Three countries conducted two case studies each—Canada, Japan, and Norway. Australia, the Netherlands, and Scotland studied the introduction of technology as a school subject into K-12 education, a topic of growing interest in the United States (Raizen et al. 1995). Another educational meaning linked to the word "technology"–issues about use of technology to provide instruction–did arise in several U.S. cases (Kids Network, Precalculus, and Voyage of the Mimi). Because technology as a school subject was not the subject of any U.S. case, however, *Bold Ventures* describes U.S. studies only as science and mathematics education, while the international report (Black and Atkin 1996) also includes technology education.

The following pages contain descriptions of case studies in other countries. The original versions of these summaries can be found in Black and Atkin (1994). We reproduce them here by special agreement.

Educational Innovations Studied

	science education	mathematics education	technology education
Australia	✔	✔	✔
Austria		✔	
Canada (2)	✔	✔	
France		✔	
Germany	✔		
Ireland	✔		
Japan (2)	✔	✔	
Netherlands			✔
Norway (2)	✔	✔	
Scotland			✔
Spain	✔		
Switzerland	✔	✔	
United States (8)	✔	✔	

AUSTRALIA (Tasmania)

Title of report	Science, Mathematics and Technology in Education (SMTE) Project
Language	English
Subject	Science, mathematics, and technology (grades K-10)
Age range	4-16 years
Nature of innovation	New curriculum content and pedagogies across the three subject areas, in response to the broad framework introduced by the adoption of the National Statements and Curriculum Profiles
Background to innovation	All six Australian states, territories and the commonwealth (nation) itself adopted in 1989 eight key learning areas around which to structure school curricula. National Statements provided a curriculum development framework; Curriculum Profiles assisted teaching and learning, offering a common language for reporting achievement.
Data— location of study	Six Tasmanian schools, two for each subject, ranging from urban primary (elementary) to rural secondary, which had undertaken innovation enthusiastically
Data— sources	A multi-site, multi-method approach which utilized data from a variety of schools in Tasmania gathered via interviews, document analysis, observations, and student journals
Key features	Whatever the subject (science, mathematics, or technology), all the schools had undertaken innovation largely in response to a perceived internal need, subsequently matched to external requirements. Student-teacher relationships have changed, with a shift to more student-centered learning. The change in curriculum content required change in pedagogy, which in turn promoted collegial interaction. A key teacher was imperative to the success of each innovation, and teachers involved in innovation were seen as more willing to take risks with their teaching. System-level authorities provided support, allowing a positive link between practice and theory for effecting reform.
Authors	Trudy Cowley and John Williamson, with Michael Dunphy
Availability of the report	Contact Professor John Williamson, School of Education, University of Tasmania, P.O. Box 1214, Launceston 7250, Tasmania. Tel: 61 03 243 288; 61 03 243 303; e-mail: john.williamson@educ.utas.edu.au

AUSTRIA

Title of report	Modern Mathematical Engineering Using Software-Assisted Approaches
Language	English (German version available)
Subject	Mathematics
Age range	15-19 years
Nature of innovation	The adoption of modern software (computer algebra, spreadsheets), for new approaches to mathematics teaching in higher technical colleges
Background to innovation	The image of mathematics has changed dramatically in recent years, with the advent of developments such as computer algebra and spreadsheets. Moreover, as the price of computer hardware continues to decline, such calculating aids become ever more accessible to education and to the students themselves. Against these developments, the Federal Ministry of Education formed a working group of specialists in the field of computer algebra to identify a modern approach to teaching mathematics. The group was to develop curriculum materials and give thought to implementation.
Data— location of study	A national survey of the teachers and students involved in the project
Data— sources	Questionnaires to the teachers and students involved, and recorded interviews with 22 of the students
Key features	Some 70 percent of mathematics teachers in higher technical colleges have asked to receive regular information about the project, along with 100 from other types of school and elsewhere. Application of software allows modeling and interpretation of data to gain new significance. Able students enjoy working with the new media and are additionally motivated, but the demands on less able students are increased. It remains to investigate more fully the effectiveness of these innovations, and it is hoped that the original group will be extended to allow participation from all over the country.
Author	Peter Schüller, Federal Ministry of Education, Centre for School Development, Vienna
Availability of the report	Contact Peter Schüller, Federal Ministry of Education, Centre for School Development, Strozzigasse 2/1, A-1080 Wien, Austria. Tel: 43 1 531 20 47 03; fax: 43 1 531 47 80; e-mail: pschuell@blackbox.or.at

CANADA (British Columbia)

Title of report   Gender-Equity in Science Instruction and Assessment—A Case
                     Study of Grade 10 Electricity in British Columbia, Canada

Language   English

Subject   Science (grade 10)

Age range   14-15 years

Nature of   The development and implementation of curriculum and assessment
innovation   activities, using contexts representing the diverse interests and back-
                     grounds of the students in an effort to promote gender equity. A particular
                     focus was an effort to make the dynamics of gender relations in the class-
                     room an explicit issue for discussion.

Background   Concern expressed and recommendations made in a number of
to innovation   provincial reports about the low female participation rate in the physi-
                     cal sciences at the secondary school level

Data—   A grade 10 class in a selective academic program within a large
location of study suburban secondary school with a multicultural student body

Data—   In-depth interviews with the 20 students in the grade 10 class and
sources   their science teacher; daily classroom observations over eight weeks of
                     instruction; student journals, assignments, and tests

Key features   An approach to gender-sensitive curriculum was developed in which
                     students had a significant role in shaping the curriculum within a frame-
                     work emphasizing social issues in science. Girls indicated that they were
                     more inclined to enroll in physics after the unit. Issues for teachers arose
                     around the pressures of time, tension between student choice and covering
                     the prescribed curriculum, gender balance in the small groups, and gen-
                     der patterns of group work. Girls tended to value assessment questions
                     that allowed them to develop answers emphasizing social relationships,
                     individual behaviors, and the environment. The girls also tended to give
                     more value than the boys to class discussions about the gender patterns
                     of group work.

Author   Jim Gaskell, University of British Columbia, Vancouver

Availability   Contact Dr. Jim Gaskell, Faculty of Education, University of British
of the report   Columbia, Vancouver, British Columbia, V6T 1Z4, Canada. Tel: 1 604
                     822 58 26; fax 1 604 822 47 14; e-mail: jimgask@unixg.ubc.ca

CANADA (Ontario)

Title of report	A Case Study of the Implementation of the Ontario Common Curriculum in Grade Nine Science and Mathematics
Language	English
Subject	Integrated mathematics and science (grade 9)
Age range	13-14 years
Nature of innovation	The implementation of a common curriculum, designed according to mandatory Ontario guidelines which specify learning outcomes, leaving content and objectives for the school to decide based on the curriculum guidelines in force
Background to innovation	Ongoing curriculum reform and concern from the later 1980s about school dropouts led to decisions that all Ontario schools would "de-stream" grade 9 and integrate science, mathematics, and technology using an outcomes-based curriculum.
Data— location of study	A medium-sized city high school which had an exemplary academic record and which pursued the intended reforms aggressively in advance of the implementation deadline
Data— sources	Investigations of students in grades 9 and 10 (the latter having experienced the reforms the previous year), and their teachers, to establish the perceptions of both
Key features	Time was made available for one teacher to develop integrated materials, but there was teacher opposition to the thematics approaches he suggested. The themes largely gave way to joining topics from existing mathematics and science courses, so that only a third of the course was integrated as the lead teacher had suggested. Teachers' unfamiliarity with subject matter was seen to be a problem, but de-streaming was mentioned as their main concern, since they did not feel prepared for it. Able students were said to be apprehensive about the slow pace of their work and less–able students had difficulty with written materials.
Authors	Barry Cowell and John Olson, Queen's University, Ontario
Availability of the report	Contact Professor John Olson, Faculty of Education, Duncan McArthur Hall, Queen's University, Kingston, Ontario K7L 3N6, Canada. Tel: 1 613 545 62 61; fax: 1 613 545 65 84; e-mail: olsonj@educ.queensu.ca

FRANCE

Title of report	The Impact of National Pupil Assessment on the Teaching Methods of Mathematics Teachers
Language	French (English version available)
Subject	Mathematics assessment at the beginning of the school year, to see how well students have mastered the skills needed to go farther with the subject
Age range	8, 11, and 15 years
Nature of innovation	The progressive implementation of procedures that give primary and secondary teachers the means for evaluation—diagnostic, training, and certification—to be used by them as they think fit. Thus, teachers are able to identify the strengths and weaknesses of their students, reflect critically on their own teaching methods, and adapt them to match the students' needs.
Background to innovation	The prevailing political object to universal secondary education is to bring 80 perent of each appropriate year-group to baccalaureate level. The Education Act of 1989 clearly defines the fundamental role of assessment in meeting this objective, first by an external report on the overall performance of the educational system and trends within it, and second by spreading an assessment philosophy through the system.
Data— location of study	(1) A national sample of teachers; and (2) the training schemes, designed to familiarize teachers with the assessment procedures, which were run by the local inspectorates and teacher training department in two académies (regional education authorities)
Data— sources	Surveys conducted on representative samples of teachers (nationally and within the two académies) to find out what they thought of the concept and methods of assessment, the benefits, and difficulties encountered. The analysis reported on the schemes and the impact of the training on the professional practices of the teachers.
Key features	Most teachers (of ages 8 and 11) believe the assessment helps them to modify their teaching, seeing it as a means to promote dialog between teachers and parents. Almost all teachers who used the assessment results reported them to parents. For teachers of age 15 students, the interest in assessment is much stronger for those following technical courses than it is in general. Teachers want more training in order to make better use of the assessment results, and the training has a greater impact when part of an ongoing process. Teachers have used the assessment procedures selectively.
Authors	Claudine Peretti, Deputy Director of Education System Evaluation, Evaluation and Planning Directorate, Ministry of Education; Clair Dupuis and Raymont Duval, ADIRM-IREM, Strasbourg
Availability of the report	Contact Claudine Peretti, Ministère de l'Education nationale, de l'Enseignement supérieur, de la Recherche et l'Insertion professionnelle, Direction de l'Evaluation et de la prospective, 142 rue du Bac, 75007 Paris France.

GERMANY

Title of report	Practicing Integration in Science Education (PING), an Innovation Project for Science Education in Germany
Language	English (German version to follow)
Subject	Integrated science (grades 5–10)
Age range	10-16 years
Nature of innovation	Collaborative research and development of an integrated school science program, with materials emphasizing the relationship between humanity and nature
Background to innovation	The establishment of secondary comprehensive schools led to their teachers wanting student-centered, integrated science education. In this, they were supported by the German National Institute for Science Education (IPN).
Data— location of study	Comprehensive schools in the state of Schleswig-Holstein, three schools (grades 5 and 6) in Brandenburg, and six Gymnasien in Thineland-Palatinate
Data— sources	Internal evaluation data from project schools, research papers, reports, protocols, other project documents, 11 interviews with teachers and administrators, to exhibit how the development and the teaching took place, and the interactions among teachers, researchers, and administrators
Key features	Collaboration among teachers, administrators, a research institute, and in-service activities allowed the development of materials that reflect the student's relationship with nature, promote responsible action, and are sensitive to the cultural aspects of the topic. An integrated science syllabus was produced for Schleswig-Holstein. There is a coordinating network that gives teachers information and access to materials; and a coordinating center for organizational development, information exchange, collection of relevant literature, revision of existing materials, and subsequent distribution. Workshops are held for teacher education and dissemination.
Authors	Henning Hansen, Rainer Buck, and Manfred Lang, Institute for Science Education, Kiel
Availability of the report	Contact Dr. K. H. Hansen, Institute for Science Education (IPN), Olshausenstrasse 62, D-20498 Kiel, Germany. Tel: 49 431 880 30 98; fax: 49 431 880 30 97; e-mail: npn27@rz.uni-kiel.de

REPUBLIC OF IRELAND

Title of report	IDEAS—A Case Study of In-Career Development in Equity and Science
Language	English
Subject	Physics and chemistry (teaching)
Age range	15-18 years
Nature of innovation	School-based in-career development organized by the Ministry of Education Inspectorate, whereby experienced science teachers were trained in the project schools on a one-to-one basis by an established teachers of physics or chemistry from another school. Other schools were associated less closely with the project. Curriculum materials have been designed and a scheme of centers developed for ongoing teacher support.
Background to innovation	The intervention project was implemented in 1985, when concern for girls' underrepresentation in physics (particularly) and in chemistry had been identified as a problem of lack of provision. Previous studies have shown that the project has successfully addressed this problem and has helped to change the aspirations of girls in the direction of scientific careers. The present study examined the question of whether the project offered a transferrable model of in-career development in equity and science.
Data— location of study	33 second-level schools (i.e., those teaching students aged 12-18), notably girls' schools or mixed schools
Data— sources	Interviews with participating teachers, questionnaire survey of school principals and all teachers in the project, statistical data on examination performance of project students
Key features	Students' performances in terminal examinations corresponded to national norms, so the project successfully achieved its objects, but curriculum development sensitive to gender equity requires more factors to be considered. The model is useful in increasing the provision of existing subjects, or where trained teachers are being reskilled to be redeployed in new areas such as technology; it was judged by participants to be excellent for in-career development.
Authors	Dearbhal Ni Chárthaigh and John O'Brien with Patricia Dundon
Availability of the report	Contact Centre for Studies in Gender and Education, University of Limerick, Plassey, Limerick, Ireland. Tel: 353 61 20 26 91; e-mail: dearbhal.nicharthaigh@ul.ie

JAPAN

Title of report	A Case Study of Teacher/Student Views About Mathematics Education in Japan
Language	Japanese (English version available)
Subject	Mathematics (grades 1-9)
Age range	6-15 years
Nature of innovation	A revised course of student to lead into the 21st century, put into effect in April 1992 for all grades in elementary school, and in 1993 for all lower secondary grades. The intention is to promote student individuality and to be responsive to international developments and to advances in technology, though with reduced teaching time available.
Background to innovation	Discussion of the revised course of study started in 1981 and was finalized in 1989. From September 1993, the second Saturday of each month became a holiday for all elementary and secondary school grades, followed from April 1995 by the fourth Saturday in addition.
Data—location of study	Six public (city or town) schools, i.e., four elementary and two junior secondary, in which classes are generally mixed, from Nara and Osaka prefectures in West Japan
Data—sources	Questionnaires and interviews for teachers and students; classroom observation and rough lesson notes; the accounts of teachers and students and others, in the school, for example, parents and education officials
Key features	Problem solving was tackled in two phases, by individual students and by students in groups supported by the teacher. Attempts were made by the teacher to anticipate student ideas and build on them positively, and to apply mathematics to other subjects and to everyday life. The approaches were designed to maintain the cognitive development of students with the reduced teaching time, with drill seen to be important. It is usual to rely on textbooks for lesson preparation.
Author	Keiichi Shigematsu, Nara University of Education
Availability of the report	Contact Professor Keiichi Shigematsu, Nara University of Education, Takabatake-Cho, Nara, 630 Japan. Tel: 81 742 27 91 84; fax: 81 742 27 91 41; e-mail: shigek@nara-edu.ac.jp

JAPAN

Title of report	Case Studies of the Implementation of a New School Science Course in Japan
Language	Japanense (English version available)
Subject	Science
Age range	9-15 years
Nature of innovation	Implementation of the revised course of study (national curriculum guidelines), in 1992 for all elementary schools and in 1993 for lower secondary schools. There are new emphases on human responsibility, individuality, and resourcefulness.
Background to innovation	The course of study is revised roughly every 10 years, and the present one was adopted in 1989. Evaluation of how it is implemented will be useful for the next revision.
Data— location of study	Six elementary schools and one lower secondary school, located in Yokohama City near Tokyo. Each school is standard and of normal size.
Data— sources	Questionnaires and interviews for students and teachers in one elementary school and the lower secondary school, records of observations and interviews in the other five elementary schools during the refinement of lesson plans before and after classes
Key features	A teacher writes a lesson plan, which is discussed with other teachers regarding the suitability of content, teaching methods, and materials. Subsequently, teachers observe the teaching-learning process during the lesson, evaluate the process afterward, and rewrite the lesson plan for the next lesson. This popular and useful method of improving lesson plans is traditionally used when a new course of study is introduced.
Authors	Toshiyuki Fukuoka, Yokohama National University
Availability of the report	Contact Professor Toshiyuki Fukuoka, Faculty of Education, Yokohama National University, 156 Tokiwadai, Hodogaya-ku, Yokohama-shi, 240 Japan. Tel: 81 45 335 14 51; fax: 81 45 333 15 36; e-mail: fukuoka@ed.ynu.ac.jp

THE NETHERLANDS

Title of report	An In-Depth Study of Technology as a School Subject in Junior Secondary Schools in the Netherlands
Language	English
Subject	Technology
Age range	12-14 years
Nature of innovation	The implementation of technology as a new subject within the basic education program, in accordance with the new national curriculum for the lower sector of junior secondary schools
Background to innovation	Government policy is to reform secondary education in the Netherlands, both for curriculum content and teaching strategies. This innovation is still continuing: It follows reform of primary education which began in 1985.
Data— location of study	Three core schools, i.e. one gymnasium (grammar), one broad-based (pre-university and pre-vocational) and one pre-vocational; additional information was collected from two comparison schools of each type
Data— sources	Seven classroom observations in each core school, supplemented by teacher interviews, four video recordings of observed lessons, journals completed by teachers and textbook analysis. In the comparison schools interviews with a member of the school management and one technology teacher at the beginning and at the end of the research period. Document analysis in all schools and a pupil questionnaire investigating pupil opinions about the technology lessons.
Key features	The gymnasium and broad-based schools emphasized knowledge, with more practical activities in the pre-vocational school. There was unresolved conflict in the nine schools between the amount of knowledge required by the curriculum and the need to develop practical skills. Teachers think cooperation with other subjects is desirable, but as yet little has been achieved. Introducing technology imposed a heavy teaching load, and the provision of satisfactory textbooks is seen to be of major importance—though as yet unmet. Assessment is seen as problematic.
Authors	Henk A.M. Franssen, Harrie M.C. Eijkelhof, Eric A.J.P. Duijmelinck, and Thoni A.M. Houtveen
Availability of the report	Contact Dr. A.A. M. Houtveen, Department of Education, P.O. Box 80140, 3508 TC, Utrecht University, Utrecht, The Netherlands. E-mail: houtveen@fsw.ruu.nl

NORWAY

Title of report	Assessment as a Link Between Instruction and Learning in Mathematics, Especially Focusing on Pupil Self-Assessment
Language	English
Subject	Mathematics (grades 7 and 8)
Age range	13-15 years
Nature of innovation	To develop a mathematics instruction and assessment practice that would stimulate students to more active participation in the learning process, strengthen their ability to reflect on their own learning, build up their belief in themselves and in their ability to utilize their own assessment and that of others in a constructive way
Background to innovation	The 1987 Norwegian curriculum guidelines saw it as fundamental that there should be equitable and suitably adapted education for all within the framework of the class, and that students should take responsibility for their own learning. Teachers have found difficulty with these principles, especially in the demanding subject of mathematics, where textbooks have a strong steering effect and all students must face the same written examination at the end of grade 9.
Data— location of study	Five representative schools from five different counties; teachers of the same grade work in teams and all mathematics teachers in the five schools have participated
Data— sources	Questionnaires for students and for teachers, teacher reports, and interviews and classroom observations
Key features	Participating schools were given different options and left free to try out what they thought would be the best means of attaining the goals. A developmental process started at the schools, whereby teachers changed their mode of teaching and assessing—some to a great extent, others not so much. Planning and information provision have been key factors. Most progress arose in those classes where teachers made plain to students and their parents the aims and methods, and sufficient time was set aside to discuss and modify plans in cooperation with the students. Some encouraging progress has been made with student self-assessment.
Author	Sigrun Jernquist, National Examination Board, Oslo
Availability of the report	Contact Sigrun Jernquist, The National Examination Board, Box 8105 Dep, 0032 Oslo, Norway. Tel: 47 22 00 38 65; fax: 47 22 00 38 91.

NORWAY

Title of report	A Case Study of Science Teaching in Grade 8, Norway
Language	English
Subject	Physics
Age range	13-14 years
Nature of innovation	Activity-based teaching methods introduced to counteract the theory-oriented teaching at this level, with the further expectation of oral examination at the end of grade 9
Background to innovation	The Department of Education had in 1992 issued guidance for oral examinations at the end of grade 9, for subjects such as science which had no national written examination. Science was seen as based on observation and experiment, so that knowledge could not be isolated from process. An oral examination based on laboratory and field experience would encourage teachers to emphasize process as well as content.
Data— location of study	A junior high school in the central part of Norway
Data— sources	Three teachers and their classes were followed throughout one term during the teaching of an electricity unit. Teachers and students were interviewed, student and teacher logs and materials collected, and observations of teaching sequences made.
Key features	The teaching materials used in the study were developed by the research team and implemented by the teachers. Teachers and students reacted positively to the approaches which were new to them, but as yet there is no knowledge of the effectiveness of the final oral examination or of transferability to other teaching themes.
Authors	Doris Jorde and Rolf Krohg Sørensen, University of Oslo
Availability of the report	Contact Dr. Doris Jorde, Department of Teacher Education and School Development, University of Oslo, PB 1099, Blindern, Oslo, Norway. Fax: 47 22 85 74 63; e-mail: doris.jorde@sls.uio.no

SCOTLAND

Title of report	A Report on Technology in Case Study Primary Schools in Scotland
Language	English
Subject	Technology
Age range	5-12 years
Nature of innovation	The implementation of the technology component of the national 5-14 guidelines for environmental studies
Background to innovation	A major review of curriculum and assessment in Scotland for pupils aged 5-14 years commenced in 1988 and led to national guidelines in identified curriculum areas. The Environmental Studies 5-14 Guidelines, published in 1993, introduced technology for the first time as a firm component of the recommended curriculum and specified the attainment levels that should be achieved by most pupils at particular stages in their schooling.
Data— location of study	Four primary schools where good practice was evident in technology innovation and a start had been made in implementing the national guidelines in technology. The schools ranged in size, location, and nature of their pupil populations.
Data— sources	Investigations of pupils in all primary years, their parents, teachers, and head teachers, to establish the experiences and perceptions of all involved, whether in planning and delivering change or in being the recipients of curriculum innovation
Key features	The four schools adopted different approaches, although in each key individuals were identified as crucial to managing change, and their roles are highlighted. Key points are identified to ensure optimum conditions are in place for implementing the introduction of technology; they are focused on education authorities, school management, and class teachers.
Authors	Peter Kormylo, head teacher of a large urban primary school, and John Frame, lecturer in science and technology at Moray House Institute of Education, Edinburgh
Availability of the report	Contact HMI Alistair F. Marquis, Her Majesty's Inspectors of Schools, Room H1-2, Saughton House, Broomhouse Drive, Edinburgh EH11 3XD, Scotland, UK. Tel: 44 131 244 84 33; fax: 44 131 244 84 24

SPAIN

Title of report	Students' Diversity and the Changes in the Science Curriculum—an Evaluation of the Spanish Reform of Lower Secondary Education
Language	Spanish (English version available)
Subject	Integrated science (lower secondary)
Age range	14-16 years
Nature of innovation	The implementation of an integrated science curriculum that uses a science-for-all approach with a focus on student diversity, which seems to be the most challenging issue for teachers and policymakers
Background to innovation	A major national educational reorganization, "the Reform," was introduced in 1990, for total adoption by 1998, but at first on a voluntary basis. The Reform for lower (as distinct from post-compulsory) secondary was organized in two cycles, namely 12-14 and 14-16, the second of which extended compulsory education by two years and is of interest in this study. The second cycle curriculum, specified in 1991, establishes nationally the minimum teaching requirements (content and general objectives) and criteria for assessment. In the Reform, science is one subject, Natural Sciences, for three years; for the fourth year, science is optional.
Data— location of study	Five schools, all within the authority of the Ministry of Education, each of which exhibits distinguishing features with regard to the Reform process. They were considered to be good schools in science, both by advisors and curriculum developers of the Ministry of Education and by local inspectors.
Data— sources	Multiple sources—including direct non-interventional on-site observation; listening; interviewing (principals, teachers, pupils, policymakers, inspectors, external advisors and others); analysis of documents and field notes—to investigate how all collaborate in fitting the curriculum to student diversity
Key features	Curriculum development leads to a closer interrelation between the school as an educational community and the classroom, where there is innovation from both teacher and student. Teachers work collaboratively in refining curriculum materials, in sequencing subject content according to levels and cycles, and in establishing criteria governing assessment and the promotion of students from one grade to the next.
Authors	Antonio J. Carretero, Juan A. Hermosa, and Maria J. Sáez Brezmes, Valladolid University
Availability of the report	Contact Professor Maria J. Sáez, Faculty of Education, C/G Hernández Pacheco 1, E-47014 Valladolid, Spain. Tel: 34 83 423 441; fax: 34 83 423 436; e-mail: maria@pinar1.csis.es

SWITZERLAND

Title of report	The Representation, the Understanding and the Mastering of Experience—Modeling and Programming in a Transdisciplinary Context
Language	English (pupil protocols and their detailed analysis, and additional material and publications available in French)
Subject	Mathematics and science
Age range	13-16 years
Nature of innovation	Integration of mathematics and science through modeling and programming, the building of a transdisciplinary space for the interaction of logico-mathematical thinking (the space of necessity) with casual thinking (the space ⌐f contingency). The construction is based on a coherent set of experiments performed by the students and on their manifold representations and models. Computer programming in LOGO provides a unified common language for the activities.
Background to innovation	The complusary secondary school curriculum of the Geneva canton has an introductory course on informatics and a course on scientific observations. This project, dependent on the availability of teachers for pedagogical research, arose in response to those requirements.
Data— location of study	Two secondary schools in the Geneva canton, with the involvement of two teachers and several classes, grades 7-9
Data— sources	Interactions between the schools and three researchers from a center for psychopedagogical studies; classroom observations and in-depth interviews with students (in pairs) for the first four years, with only classroom activity in the fifth year; discussion with teachers
Key features	The two teachers were freed for one afternoon a week to hold meetings with the researchers, and in school implemented the project within their own classes. Students have reached some understanding of the natural phenomena of growth and change and the ways in which patterns can be established and generalized, but it is not yet established how the work can be extended to other schools and other curriculum areas.
Authors	Claude Béguin, Martial Denzler, Olivier de Marcellus, Anastasia Tryphon, and Bruno Vitale
Availability of the report	Contact (and for other published papers) Bruno Vitale, 27 Gares, 1201 Geneva, Switzerland. Tel: 41 22 733 52 11.

242

NAME INDEX